T0241796

System Engineering Approach to Planning Anticancer Therapies

Andrzej Świerniak • Marek Kimmel •
Jaroslaw Smieja • Krzysztof Puszynski •
Krzysztof Psiuk-Maksymowicz

System Engineering Approach to Planning Anticancer Therapies

 Springer

Andrzej Świerniak
Institute of Automatic Control
Silesian University of Technology
Gliwice, Poland

Marek Kimmel
Rice University
Houston, TX, USA

Institute of Automatic Control
Silesian University of Technology
Gliwice, Poland

Jaroslaw Smieja
Institute of Automatic Control
Silesian University of Technology
Gliwice, Poland

Krzysztof Puszynski
Institute of Automatic Control
Silesian University of Technology
Gliwice, Poland

Krzysztof Psiuk-Maksymowicz
Institute of Automatic Control
Silesian University of Technology
Gliwice, Poland

ISBN 978-3-319-80270-1   ISBN 978-3-319-28095-0   (eBook)
DOI 10.1007/978-3-319-28095-0

Mathematics Subject Classification (2010): 93A30; 92C42, 49N90, 92D25, 37N25, 92C50, 93C95, 35Q92, 93B35.

© Springer International Publishing Switzerland 2016
Softcover reprint of the hardcover 1st edition 2016
This work is subject to copyright. All rights are reserved by the Publisher, whether the whole or part of the material is concerned, specifically the rights of translation, reprinting, reuse of illustrations, recitation, broadcasting, reproduction on microfilms or in any other physical way, and transmission or information storage and retrieval, electronic adaptation, computer software, or by similar or dissimilar methodology now known or hereafter developed.
The use of general descriptive names, registered names, trademarks, service marks, etc. in this publication does not imply, even in the absence of a specific statement, that such names are exempt from the relevant protective laws and regulations and therefore free for general use.
The publisher, the authors and the editors are safe to assume that the advice and information in this book are believed to be true and accurate at the date of publication. Neither the publisher nor the authors or the editors give a warranty, express or implied, with respect to the material contained herein or for any errors or omissions that may have been made.

Printed on acid-free paper

This Springer imprint is published by Springer Nature
The registered company is Springer International Publishing AG Switzerland

Preface

This monograph is destined for researchers working in the area of computational systems biology, design of chemotherapy and radiotherapy protocols, and related areas, as well as for graduate students and postdoctoral researchers. One distinctive feature of the book is its interdisciplinary nature, so it will be useful for readers with mathematical or biological and medical background.

We are not aware of any book that deals with exactly the same topic. There is a large number of books in systems biology, several monographs devoted to computational or mathematical oncology, and of course a huge literature on mathematical modeling in cell and molecular biology and on control theory. However, none of these approaches is sufficient for our purposes.

We would like to thank biologists, with whom we have a longstanding cooperation, Profs. Joanna Rzeszowska and Maria Widel from the Institute of Automatic Control, Silesian University of Technology, and Piotr Widlak (with his group) from Maria Sklodowska-Curie Memorial Cancer Center and Institute of Oncology Gliwice Branch for their valuable insights into molecular biology and oncology and their help in interpretation of modeling results. Profs. Barbara Jarzab and Rafal Tarnawski from Maria Sklodowska-Curie Center shared with us their knowledge on oncology and clinical practice. We would like to acknowledge also our cooperation with Allan Brasier from University of Texas, Medical Branch, Galveston, Texas, USA, particularly in the area of research on signaling pathways.

This book has been written with the support of the grant of the Polish National Science Centre, DEC-2012/04/A/ST7/00353 (MK, AS, JS), and was partially supported by the grants DEC-2012/05/D/ST7/02072 (KP) and DEC-2011/03/B/ST6/04384 (KPM).

Gliwice, Poland   Andrzej Świerniak
Houston, TX, USA, Gliwice, Poland   Marek Kimmel
Gliwice, Poland   Jaroslaw Smieja
Gliwice, Poland   Krzysztof Puszynski
Gliwice, Poland   Krzysztof Psiuk-Maksymowicz
September 2015

Contents

Chapter 1
Introduction

Abstract The book shows the modeling cancer growth and cancer therapy from system engineering point of view. It might be an approach different from those most often used. Therefore, the Introduction section is devoted to explaining this perspective. It also presents a brief review of issues associated with modeling treatment of diseases and general outline of book contents.

System engineering approach in cancer research has a long history (see, e.g., [4, 16] for the survey of early research in this area) , although it seems that until recently it has been under-appreciated relative to its contributions and influence. Feedback control of biological populations has been stressed and discussed by the early classics of systems theory such as von Bertalanffy and Wiener. The first biomathematical journal "Mathematical Biosciences" has been founded by Richard Bellmann, the inventor of dynamic programming, which under various guises such as the Viterbi algorithm has been one of the most fruitful and commonly employed approaches to recursive optimization and estimation (enough to mention the hidden Markov models). The 1970s and 1980s were the period in which high hopes were attached to systems approach. This is the period in which the clonal resistance theory was announced by Goldie and Coldman and when Moolgavkar–Knudson model of carcinogenesis was formulated. A number of models of cytotoxic action of chemical agents and radiation have been tested in vitro and some of them in experimental animals. This initial enthusiasm has subsided in the following decade for a variety of reasons. One of them was that the system approach did not fit the hypothesis driven research paradigm that has dominated (and still to some extent does) among biologists. Another was a traditionally low average level of mathematical literacy among biologists and physicians (which currently is undergoing a fundamental change). Still another important reason was that parameters of the cell proliferation systems in human body were non-measurable. In addition, a fundamental feature of spontaneous human cancers, their bewildering heterogeneity, could not be gauged using mostly deterministic ODE models of treatment. On the other hand, the period of late 1990s and early 2000s (these benchmarks being very approximate) is one of the explosive growth of bioinformatics and its achievements such as human genome sequencing, sequence annotation, as well as high-tech methods of retrieval of information from individual cells, visualization of tissues, and even millimeter-size

© Springer International Publishing Switzerland 2016
A. Świerniak et al., *System Engineering Approach to Planning Anticancer Therapies*, DOI 10.1007/978-3-319-28095-0_1

tumors. Gradually, progress in these technologies allows building models which are more realistic and not only allow understanding failures of cancer treatments but also hold promise of prognosing progression and optimize treatment. Many of these new approaches are related to the "hallmarks of cancer" as listed in the two seminal papers by Hanahan and Weinberg. This includes realization of importance of vascularization in solid tumors, and more generally recruitment of normal cells by tumors. Another important element is the understanding of the role of altered metabolism in cancer cells. Still another is consideration of existence of cancer stem cells and their potentially high drug resistance as well as plasticity.

The essence of the systems engineering approach is: First to describe cancer disease as a dynamical system, whose properties represented by state variables change in time, by means of mathematical formulae. Second, to introduce anticancer therapy as a control action on this dynamical system and to search for controls that are feasible and satisfy the therapy goal.

The approach described here contributed to the development of ideas of chemotherapy scheduling, multidrug protocols, and recruitment and synchronization of cancer and normal cells. It also helped in refinement of mathematical tools of control theory applied to the dynamics of cell populations. However, few of these studies found way into clinical application. The reasons for this failure are that, on one hand, important biological processes were unrecognized and crucial parameters were not known and, on the other, that the mathematical intricacy of the models was not understood. Yet another reason for this failure might have been in the way the results of mathematical modeling have been, and still are, presented. Only a handful of papers (e.g., [1, 9]) present what catches the attention of clinicians, which is either so-called Kaplan–Meier survival plots or event-free survival plots. It seems to be a necessary element of modeling, as these plots represent the ultimate clinical result that can be compared to existing clinical studies.

On the other hand, even existing modeling studies provide clear indications about how to improve existing therapy protocols. For example, it was found in [13] that weekend gaps in radiotherapy might have enormous, unnecessary impact on treatment efficacy. Unfortunately, conclusions of these findings have never been implemented due to organizational impediments in the clinics.

The situation is changing rapidly because of the development of a new science of systems biology which builds bridges over the gap between mathematicians and engineers on one side and biologists and clinicians on the other.

Before various models can be introduced and methods to analyze them explained, we begin with introducing the concepts of a *dynamical system* and *control* in the context of cancer modeling.

In classical control theory (and system engineering) we consider a dynamical system (a system, whose behavior changes in time), called *a plant*, and the goal is to design the so-called *control law*, i.e. a way in which a variable or variables, called control, affect the system so as to obtain desired system behavior. This could be achieved in so-called open loop control structure (see Fig. 1.1a), if the exact model is known and there are no disturbances that affect plant dynamics or control itself. If applied to biological systems, the plant in this scheme represents a whole

Fig. 1.1 (**a**) Classical open loop structure of a control system; (**b**) system theoretic view of a therapy; (**c**) system in which cancer cells population is an object of control

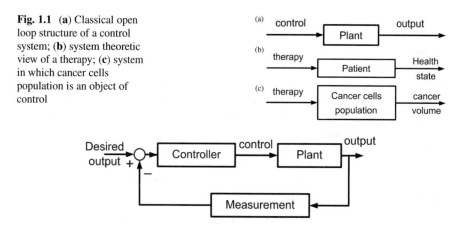

Fig. 1.2 A general diagram of a closed-loop control system

organism, tissue, population of cells or individual cell, depending on what is under consideration. For example, if the investigation is focused on a patient, his/her body is the plant, the therapy constitutes control (Fig. 1.1b), and the output of the system is the state of the patient. Such general description, however, is not precise enough for design purposes. Instead, a simpler model would be recommended to start with. Then, a population (or populations) of cells is the object of control, which, in turn, represents therapeutic actions. Accordingly, the population size can be regarded as the output of the system that should be driven to the neighborhood of zero (or below a given level) by the control (Fig. 1.1c).

Since the assumptions of disturbance-free environment and flawless model with accurately specified parameters are unrealistic, another system structure is preferred, the so-called closed-loop control system (Fig. 1.2), containing a negative feedback. Its form allows compensating uncertainty of the model parameters or conditions in which the system works. Unfortunately, although preferable from the point of view of its properties, the closed-loop control rarely can be used in biological or medical applications. Development of closed-loop systems that can be applied to cure a disease, and, in case when it is impossible, to support more or less normal life of a patient, requires:

- knowledge of physiological mechanisms underlying the illness;
- technology for measuring important biological markers that provide information needed for control;
- non-invasive measurement techniques;
- technology for providing the therapeutic input.

Therefore, though such regulation circuits exist, e.g. in glucose blood level control [3, 15], respiratory control [6], artificial kidney [5, 8], and other areas of technological support in medicine [2], in cancer they are not feasible, due to lack of effective measurement techniques. This is the reason why, at least in the near future, only the open loop therapy appears feasible in the field of cancer treatment. Moreover, contrary to the standard control theory applications in which the control time horizon is often infinite, in the cases discussed in this book it has to be finite and usually is predetermined. Accordingly, the models and methods in this book are chosen with this viewpoint in mind.

No matter what control system structure has been chosen, the problem statement must define the ultimate goal to be achieved using control. Roughly speaking, we might be interested in one of the following aims:

- eradication of cancer cells or driving cancer population below a desired threshold level;
- finding a therapy optimal in the sense of a particular performance index.

The first goal can be achieved using constant or periodic therapies that can be found mathematically. The drawback of any approach to achieve this goal, is a long (in mathematical terms even infinite) therapy period. On the other hand, the second goal makes possible to limit the duration of the treatment protocols and cumulative dose of the drugs, which is necessary because of availability of the drugs, their costs and side effects. Having accepted such framework, the next step is to define the search for drug administration protocols in the terms of an optimization problem. Such approach differs from models which merely include therapy through changes in parameters, even though some of these latter are successful in matching experimental data (e.g., [7, 10, 11]). The control system approach, particularly when it is combined with pharmacokinetics (PK) and pharmacodynamics (PD) models, allows finding an optimal schedule and dose of an anticancer (or antiangiogenic) drug.

One acceptable solution is to formulate optimal control problem for the system given by the chosen model and the control objective which adequately represents the primary goal of the therapy, which in clinical practice is maximization of tumor cure probability (TCP) that is proportional to [14]:

$$\text{TCP} = \exp\left(-f \cdot \theta \cdot N(T_k)\right), \tag{1.1}$$

where f, θ, and $N(T_k)$ denote clonogenic fraction, tumor cell density, and the number of cancer cells at the end of therapy, respectively. T_k represents the time horizon of the therapy. The number of cells is affected directly by the therapy, so the index to be minimized may be defined by

$$J = N(T_k). \tag{1.2}$$

However, the total dose of drug (or radiation, in the case of radiotherapy) has to be bounded and its instantaneous value should be constrained. This leads to control optimization problems with constraints, to be discussed in the book.

The role and use of control theory in planning anticancer therapies was recognized almost 40 years ago (see [12] for the survey in this area). We may distinguish three levels at which anticancer therapy can be considered as a control problem:

- modeling cell cycle as an object of control under action of anticancer chemotherapy including cell cycle phase dependence of the drugs and drug resistance of neoplastic cells;
- using nonlinear models of population dynamics to design protocols of antiangiogenic, immune, and combined anticancer therapies;
- modeling therapies targeting specific cellular regulatory networks important for carcinogenesis, cancer growth, and development.

The first two approaches have been discussed in detail in the literature. They involve cooperation of scientists from engineering and mathematical community and cancer research institutions. The third approach is now one of the hot spots of mathematical oncology.

In this book we want to show that system engineering approach can be useful in analysis of different therapy approaches and complement experiment in expanding knowledge of processes that must be considered when dealing with cancer. Even more, it is a necessary tool for designing therapy protocols that are directly used in medicine.

This results from a long-term interest in modeling control of cancer cell populations and resulting mathematics and also by participation of the authors in experimental and clinical work. The authors are affiliated with the Systems Engineering group at the Silesian University of Technology (SUT) in Gliwice, Poland and one of them also with the Departments of Statistics and Bioengineering at Rice University in Houston, TX, USA. The path leading to this publication started when around year 1978, Andrzej Swierniak and Marek Kimmel became interested in mathematical modeling of the cell cycle and used their background in applied mathematics and control theory to state and solve very idealized problems of optimization of chemotherapy protocols (some of them published in the Technical Reports of the SUT). Already at this point they became involved in collaboration with the Institute of Oncology (IO) in Gliwice, first with Professors Chorazy and Sznajberg and then with many others. This collaboration has become more difficult when in 1982, Kimmel moved to New York and started working at the Memorial Sloan-Kettering Cancer Center, among others on modeling of experiments concerning not only cell cycle-specific action of cytotoxic compounds. This work resulted in insights concerning estimation of parameters of the models but also an understanding of mechanics of the cell cycle. In the meantime, Swierniak continued work on optimization problems, mostly in the framework of ordinary differential equation (ODE) models. In the early 1990s, Kimmel moved to Rice University and started working with a biologist David Axelrod at Rutgers on probabilistic models of genome evolution (specifically gene amplification) and Swierniak, now aided by Jaroslaw Smieja and Andrzej Polanski, ventured in the direction of infinite-dimensional control problems, partly based on Kimmel–Axelrod models. The period after year 2000 was one of an explosive growth of bioinformatics, genomics, computational biology, and finally systems biology. This has been reflected in growth of these disciplines at the SUT, where an interdepartmental Center for Biotechnology with undergraduate and graduate programs has been created. This was accompanied by a vigorous expansion of research activities partly under the umbrella of the EU-funded Biopharma Center, and exchange of resources and personnel between SUT and IO. Inevitably, interests (computational and experimental) of now numerous researchers started diverging. However, the optimal control thread persisted in collaboration of mathematicians Urszula Ledzewicz and Heinz Schaettler, and many others. One of the milestones was participation of Kimmel, Swierniak, and Smieja in the activities of the NSF-funded Mathematical Biosciences Institute in Columbus, OH, USA and resulting new contacts. Among authors of the current book are also representatives of the new generation, Krzysztof Puszynski whose field is modeling of drug action on signaling pathways in cells and Krzysztof Psiuk-Maksymowicz specializing in spatial models of tumor growth and therapy.

The scope of this book includes a range of models of evolution of cancer cell populations, analysis of their dynamics, modeling various types of anticancer therapy, as well as introduction into modeling of intracellular processes associated with carcinogenesis, intercellular communication, and therapy-induced processes.

Since the efficacy of anticancer treatments is related to the effects they induce in cancer and normal cells and these, in turn, depend on cell cycle, we start with modeling of cell cycle. Sophisticated models that take into account many intracellular processes are not well suited to the task of designing control strategies. Compartmental models are more convenient if the goal is to build approximate and manageable models. They will be discussed in Chap. 2, in which we will also mathematically formulate the optimization problem. Solution of this problem will lead to recommendations for therapy protocols that are valid for most chemotherapy and combined therapy protocols. We will then consider pharmacokinetics and pharmacodynamics of drugs, as well as other effects including resonances, synchronization, and aftereffects. Subsequently, we will deal with drug resistance in chemotherapy, its various biological sources and methods to overcome it.

Chapter 3 is focused on population dynamics approach to therapy optimization. We consider cells populations there, without distinguishing phases of the cell cycle in which the cells are. This approach is particularly valid, when describing antiangiogenic and combined therapies, as well as immunotherapy. Several examples of such treatment are introduced in the chapter and discussed with respect to their dynamical properties and optimization of treatment protocols.

The population-centered point of view, or even the cell cycle-centered point of view, are approaches that constitute the first steps in understanding the intricacies of the problem at hand. Therefore, in Chap. 4 we describe structured models and their use in modeling of anticancer therapies. Spatial models of cancer growth are discussed, as well as age- and physiologically structured models. The methods that are applied to analyze cancer growth and its inhibition using therapy also involve cellular automata and these latter are covered in the chapter as well.

Discussion presented in Chaps. 2–4 shows that to be effectively applied, control theory-based analysis must be preceded by broadening basic knowledge of intracellular processes contributing to cancer development, cancer resistance to a therapy and, finally, efficacy of the therapy chosen. This leads to Chap. 5, in which intracellular processes are discussed. Deterministic and stochastic models are presented, supported by sensitivity analysis that can indicate prospective molecular targets in new therapies. Use of interfering RNA in cancer treatment is discussed at the end of the chapter.

Quantitative recommendations for therapy protocols have to be based on models that are quantitative, and include biologically valid parameters. Chapter 6 deals with model identification and parameter estimation.

The book concludes with the discussion of applicability of the models and perspectives of their development and clinical importance. Discussion chapter is more philosophical in nature and does not involve any equations. It is recommended as initial reading for those new to the field.

Four Appendices are found at the end of the book, containing important mathematical methods, tools, and references needed in analysis of the problems introduced in this book, as well as many others related to biomathematics.

Although the models presented in this book were developed to describe various aspects of cancer evolution and cancer therapy, their form is general enough to be used in other applications, where a population of cells or virions is under consideration. This includes, for example, modeling of viral or bacterial infections. For consistency, notation in some of the models presented has been altered compared to the original publications.

References

1. L. Barazzuol, N.G. Burnet, R. Jena, B. Jones, S.J. Jefferies, N.F. Kirkby, A mathematical model of brain tumour response to radiotherapy and chemotherapy considering radiobiological aspects. J. Theor. Biol. **262**(3), 553–565 (2010)
2. C. Cobelli, E. Carson, *Introduction to Modeling in Physiology and Medicine* (Academic, Amsterdam, 2008)
3. C. Cobelli, C.D. Man, G. Sparacino, L. Magni, G. De Nicolao, B. Kovatchev, Diabetes: models, signals, and control. IEEE Rev. Biomed. Eng. **2**, 54–96 (2009)
4. M. Eisen, *Mathematical Models in Cell Biology and Cancer Chemotherapy*. Lecture Notes in Biomathematics, vol. 30 (Springer, New York, 1979)
5. V. Gura, C. Ronco, A. Davenport, The wearable artificial kidney, why and how: from holy grail to reality. Semin. Dial. **22**(1), 13–17 (2009)
6. M.C.K. Khoo, *Physiological Control Systems* (IEEE, New York, 2000)
7. P. Kim, P. Lee, D. Levy, Modeling imatinib-treated chronic myelogenous leukemia: reducing the complexity of agent-based models. Bull. Math. Biol. **70**, 728–744 (2008)

8. J.C. Kim, F. Garzotto, F. Nalesso, D. Cruz, J.H. Kim, E. Kang, H.C. Kim, C. Ronco, A wearable artificial kidney: technical requirements and potential solutions. Expert Rev. Med. Devices **8**(5), 567–579 (2011)

9. N.F. Kirkby, S.J. Jefferies, R. Jena, N.G. Burnet, A mathematical model of the treatment and survival of patients with high-grade brain tumours. J. Theor. Biol. **245**, 112–124 (2007)

10. N. Komarova, D. Wodarz, Drug resistance in cancer: principles of emergence and prevention. Proc. Natl. Acad. Sci. **102**(27), 9714–9719 (2005)

11. F. Michor, T.P. Hughes, Y. Iwasa, S. Branford, N.P. Shah, C.L. Sawyers, M.A. Nowak, Dynamics of chronic myeloid leukaemia. Nature **435**, 1267–1270 (2005)

12. A. Swierniak, M. Kimmel, J. Smieja, Mathematical modeling as a tool for planning anticancer therapy. Eur. J. Pharmacol. **625**(1–3), 108–121 (2009)

13. R. Tarnawski, J. Fowler, K. Skladowski, A. Swierniak, R. Suwiski, B. Maciejewski, A. Wygoda, How fast is repopulation of tumor cell during the treatment gaps? Int. J. Radiat. Oncol. Biol. Phys. **54**, 229–236 (2002)

14. S.L. Tucker, H.D. Thames, J.M.G. Taylor, How well is the probability of tumor cure after fractionated irradiation described by Poisson statistics? Radiat. Res. **124**, 273–282 (1990)

15. S.A. Weinzimer, G.M. Steil, K.L. Swan, J. Dziura, N. Kurtz, W.V. Tamborlane, Fully automated closed-loop insulin delivery versus semiautomated hybrid control in pediatric patients with type 1 diabetes using an artificial pancreas. Diabetes Care **31**(5), 934–939 (2008)

16. T.E. Wheldon, *Mathematical Models in Cancer Chemotherapy*. Medical Sciences Series (Hilger, Bristol, 1988)

Chapter 2
Cell Cycle as an Object of Control

Abstract This chapter is devoted to models of cancer growth and anticancer therapies that put special emphasis on the dependence of therapy efficiency on cell cycle. First, biological background is introduced and detailed description of a cell cycle is given, based on the review of biological literature. It is supplemented with information about chosen chemotherapeutic drugs and their efficacy with respect to the cell cycle. Next, pharmacokinetic and pharmacodynamics aspects of chemotherapeutics are briefly described. They along with cell cycle specificity of drugs may lead to various phenomena of resonances and aftereffects that need to be taken into account in therapy and synchronization of treatment protocols. These issues are mentioned in a separate section of this chapter. Finally, models that incorporate evolution of drug resistance are presented. For all models, the problem of finding a suitable treatment protocol is formulated as a problem of control optimization and some results of application of optimization theory to solve these problems are presented.

2.1 Cell Cycle and Chemotherapy

Carcinogenesis results from loss of control over the cell cycle leading to an altered behavior of a part of cell population. Treatments directed against cancer may be viewed as control actions for which the cell cycle is the object [109, 111] Control models are formulated to enable analysis and optimization of the diverse dynamic protocols of drug administration.

Cell cycle is a sequence of phases traversed by each cell from its birth to division. These phases are: G_1—growth; S—DNA synthesis; G_2—preparation for division (gap phase); and M—division. After division, the two progeny cells usually reenter G_1. However one or both progeny may become dormant or resting, or in other words, enter the quiescent G_0 phase. From G_0, after a variable and usually long time, the cells may reenter G_1 ([5], pp. 984–987, see also Fig. 2.1). Multiple regulatory mechanisms help to maintain the cell cycle and, ultimately, lead to cell division after which two cells with the same genetic information begin their life cycles. While there is a lot of redundancies in these mechanisms, when several of their components fail, the errors may aggregate, propagate through the intracellular and

© Springer International Publishing Switzerland 2016
A. Świerniak et al., *System Engineering Approach to Planning
Anticancer Therapies*, DOI 10.1007/978-3-319-28095-0_2

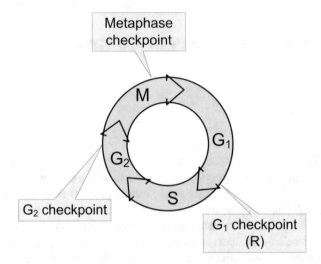

Fig. 2.1 Cell cycle. The phases and checkpoints illustrated here are described in the text

intercellular signaling networks signaling pathways leading to the development of cancer [20].

Cell populations, with a very few exceptions, are not synchronized with respect to the phases of the cell cycle of individual cells are in. This is one of the most important issues in modeling cell growth and responses to therapeutic actions. In an equilibrium population, the fraction of cells being in a particular phase is related to the ratio of the length of this phase and the total duration of the cell cycle. This property is usually used to help to determine the average length of the cell cycle.

Cell cycle is controlled through a cascade of biochemical events, such as phosphorylation and dephosphorylation processes, which are catalyzed by enzymes called kinases and phosphatases, respectively. Kinases that are at the core of cycle regulation are called cyclin-dependent kinases (Cdks) and are activated by a family of cyclins whose members are associated with transitions from one phase to another [5, p. 997]. Knowledge about these processes is important, if the models are to include intracellular processes to be targeted by a therapy.

For each tissue it is possible to compute its mitotic index, i.e. fraction of cells in the M phase. Though its value fluctuates due to various factors, in normal conditions it can be treated as constant, resulting in mitotic homeostasis. Through a balance with cell turnover, the number of cells in organs of an adult is kept constant, which is an indicator of a healthy state. Moreover, most of the cells remain in the quiescent G_0 phase. They can enter the cell cycle if activated by a range of external signals, such as growth factors, neuromediators, hormones, or by direct interaction between neighboring cells or between cells and the extracellular matrix [29]. Most of these signals enter the cell through cell membrane receptors.

Understanding the biological background of the cell cycle is extremely important, as it forms the basis for biologically acceptable models.

A cell entering G_1 phase is approximately one half of the size of its parent cell. The length of this phase is the most variable, compared to other phases and ranges from several to more than 10 h in human cells. Accordingly, the lengths of its subphases also exhibit large variability within the cell population.

The G_1 phase is characterized by intensive anabolic processes, substantial chemical exchange with extracellular environment, as well as other types of activity, such as motility, membrane transport, and so forth. During this phase macromolecular content of the cell increases, which leads to increase of cellular mass and volume. In the early G_1 phase, the cell reaches the so-called checkpoint R and if it is passed, replication of DNA is initiated. If the checkpoint R for any reason is not passed, the cell enters the quiescent phase G_0. Mechanisms of checkpoint transition have been uncovered only recently and the complete regulatory mechanism is still unknown. Most likely, it is associated with the level of regulatory proteins such as Cdc6 that have relatively short half-life times, and their phosphorylation.

Cells in the G_1 phase, or, more specifically, on the boundary between G_1 and S phase, can be arrested both in vivo and in vitro by pharmacological agents, including, among others, 5-fluorouracil or methotrexate (though these two are mainly used as killing agents or for blocking DNA synthesis, resulting in the arrest in the S phase), which inhibit biosynthesis of endogenous purine or pyrimidine bases [4, pp. 71–77]. This leads to a $G_1 - S$ block, resulting in synchronization of cells in the late G_1 phase. This block is utilized in some chemotherapy protocols.

Before each cell division, the amount of DNA in the cell nucleus doubles so that each progenitor cells obtains full DNA content. Doubling of DNA takes place in the S phase that takes approximately 7 h in mammalian cells. Again, the mass and volume of a cell must increase, compared to the G_1 phase. Similarly, as in the case of G_1 phase, cells can be arrested in the S phase both in vivo and in vitro by means of agents such as mitomycin C [4], leading to synchronization of cell cycle.

Regulation of the S phase consists in controlling the $G_1 \rightarrow S$ transition and checking if DNA synthesis is complete. Molecular mechanisms of DNA replication checkpoint are unknown [56]. However, it seems safe to assume that these mechanisms prohibit repeated synthesis of replicated DNA. It is speculated that p34-cyclin takes part in this process, though its level is kept constant throughout the cell cycle [5]. The mathematical model of DNA replication can be found in reference [55].

The G_2 phase is a gap phase between the end of the S phase and initiation of mitosis. Its duration in mammalian cells is usually 1–2 h. One of the important processes taking place in this phase is production of the proteins that form the mitotic spindle, components needed for recreation of the nuclear envelope and actin and myosin filaments that are needed in the telophase and cytokinesis subphases of mitosis.

Duration of the G_2 phase depends on several environmental factors. Lowered temperature decreases the frequency of oscillatory movement of the cell membrane and leads to elongation of the G_2 phase. Ionizing radiation can also lead to cell arrest in this phase and its effects are stronger than in the preceding phases.

The *M* phase consists of the nuclear division, called mitosis, and the cytoplasmic division, called cytokinesis. Mitosis itself is divided into five subphases: prophase, prometaphase, metaphase, anaphase, and telophase, which can be distinguished by spatial distribution of chromosomes, centrosomes, and spindle poles.

The prophase begins with initiation of chromatin condensation, which leads to creation of mitotic chromosomes. Each of them consists of two sister chromatids, held together by protein complexes called cohesins and linked at the centromere location. Later, special protein complexes called kinetochores form on each centromere and subsequently serve as attachment places for mitotic spindle microtubule. Chromatin condensation continues up to the very end of the prophase and results in creation of fully condensed chromosomes in the metaphase.

Other processes taking place at the prophase enable effective mitosis possible. In the cytoplasm, the cytoskeleton microtubules are decomposed into separate tubulins. These along with the tubulins produced in the G_2 phase are used to build the mitotic spindle—a bipolar structure, consisting of microtubules, dyneins, and microtubule associated proteins (MAPs). The mitotic spindle is at the core of mitotic machinery that includes also other elements that allow it to connect to intermediate filaments and thus stabilize its position. As one of the elements most crucial for effective mitosis, this mitotic spindle has become a molecular target for various cytotoxic agents.

In fact, first elements of the mitotic spindle are starting to form as early as in the late G_1 phase. Then, two centrioles which up to certain point have been close to each other separate by a few micrometers and the process of their duplication begins. As a result, two pairs of centrioles are created. In the prophase the two pairs separate from each other and form the poles of the mitotic spindle that subsequently increase the distance between them. Each centriole pair is associated with its own array of microtubules called an aster and kinetochore microtubules. The structure of these microtubules undergoes substantial changes during mitosis, promoting relocation of chromosomes in later subphases.

An important characteristics of the prophase is substantial decrease of biosynthesis of both proteins and RNA, as well as carbohydrates. This results from reduced transcriptional activity due to chromatin condensation.

Prometaphase begins with the breakdown of the nuclear envelope. It is triggered by M-Cdk kinase, which phosphorylates the nuclear lamina that underlies the nuclear envelope, changing its spatial conformation. As a result, the microtubules of the mitotic spindle gain access to the condensed chromosomes. A complex relocation process then begins, at the end of which (at the metaphase) the chromosomes assume a position equidistant between the two spindle poles, called a metaphase plate.

At metaphase the chromosomes are condensed even further and this process is ended when all of them take their equatorial position. Though their location is now specific, they remain in a dynamic equilibrium, continuously oscillating around the equilibrium location. At the same time cytoplasmic content move towards the poles so that to provide each progeny with approximately half of it.

At anaphase sister chromatids separate at the centromere locations. The kinetochore microtubules contract pulling daughter chromosomes towards the spindle poles. Moreover, the spindle poles also move apart, thus leading to chromosome separation. It is speculated that it is the centromere DNA synthesis, taking place at the beginning of anaphase that allows breaking the links between sister chromatids (as the centromere DNA is not synthesized in the S phase and thus centromeres of sister chromatids are merged). Usually, all chromosomes are relocated at the same speed, which is inversely proportional to the distance between chromosomes and spindle poles (i.e., it decreases as they get closer to each other).

The mechanism of chromosomes relocation at prometaphase, metaphase, and anaphase is not fully understood. However, it must be based on interactions between microtubules, as well as elongation of mitotic spindle that results in elongation of the microtubules. Therefore, when considering actions directed against mitotic cell division, substances that exert their effects on the microtubules seem to be natural choice. Indeed, substances such as vinca alkaloids (e.g., vincristine, vinblastine) are tubulin polymerization inhibitors (or microtubule-destabilizing class of antimitotic drugs). Through depolymerization of microtubules and destroying mitotic spindles they block cancer cells in metaphase, as lack of microtubules makes relocation of chromosomes towards the poles impossible. Such phenomenon is called statmokinesis and is used to synchronize cell population in mitosis. Taxanes (e.g., paclitaxel or docetaxel), on the other hand, inhibit depolymerization, disrupting reorganization of the microtubule network required for mitosis and cell proliferation [87, pp. 79–82]. These substances are among these most often used in cancer chemotherapy and belong to the group of antimitotic cytotoxic drugs (spindle poisons).

During telophase the two sets of daughter chromosomes arrive at the poles of a spindle and the process of their decondensation begins. At the same time, due to decondensed chromatin rRNA is rapidly synthesized and a new nuclear envelope is formed around each set of chromosomes, leading to restoration of the nuclei. Mitotic spindle is disassembled and part of its components is used to build cytoskeleton typical for a cell at interphase.

Cell division is concluded after cytokinesis, the process in which cytoplasm is divided into two. In fact, contrary to the previously described subphases of mitosis that take place in strict sequential order, cytokinesis does not constitute a separate subphase but spans from anaphase till the final act of creating two daughter cells.

The G_0 is called the quiescent phase. Cells in G_0 perform their function but do not have the ability to replicate genetic material and divide. The length of this phase ranges from several days to several months and longer. Most cells of the adult organs such as pancreas, liver, kidney are not in the cell cycle and virtually do not divide (doubling times are measured in years). However, after removing a part of the organ, in the case of oxygen deprivation or as effects of chemical compounds such as folic acid, isoproterenol, they are recruited back into G_1 phase [4]. Cells of many types reenter cell cycle after being affected by growth factors or hormones. This results in activation of a cascade of processes mediated by various kinases that involves the following steps:

1. Heterogeneous nuclear RNA (hnRNA) and ribosomal (rRNA) synthesis,
2. Synthesis of enzymatic and structural proteins,
3. Synthesis of DNA and histones,
4. Synthesis of the mitotic spindle proteins and mitosis.

Most cells enter the G_0 phase from G_1, however some cells (e.g., epidermal) may enter it from G_2 [5, pp. 1015–1016]. Usually, the longer cells remain in G_0, the longer it takes them to reenter the cell cycle.

Existence of G_0 cells and their ability to reenter the cell cycle is important when cytostatic-based anticancer therapy is considered. Such cells exist in most cancer populations and are insensitive to cytostatic agents. Therefore, if they reenter cell cycle after the therapy is completed, cancer growth is reactivated.

In solid tumors, there exists a geometric gradient of oxygen and nutrients. This causes stratification in viability of cells. Usually, cycling cells are located near the surface or near blood vessels; more distant layers are occupied by dormant cells, while the remotest regions form a necrotic core. This may lead to self-limiting growth phenomena, which may be described by nonlinear differential equation models (e.g., [51]). Although such models have been used to model the growth of cancer cell population therapy scheduling, they do not include cell cycle structure. They will be considered by us in Chap. 3.

The feature, which is particularly interesting for modeling of chemotherapy scheduling, is cell cycle specificity of majority of anticancer agents. This feature is essential for the initial period of chemotherapy, when at issue is the most efficient reduction of the cancer burden. This is particularly important in leukemias but may be also of interest in solid tumors. Cell cycle specificity of cytotoxic drugs is important since it makes sense to apply anticancer drugs when cells gather in the sensitive phases of the cell cycle. Therefore, the models used to describe drug actions should make it possible to distinguish cells in various cell cycle phases. Compartmental modeling is a convenient approach to do that. The idea is to group cells into subpopulations, representing cell cycle phases. Usually, cells in G_2 and M phases are grouped together in one compartment G_2–M. This is justified, among other, by the fact that parameters representing transit times between these two phases are difficult to obtain, as it is difficult to precisely distinguish cells in G_2 from those in M phase, using techniques such as DNA staining.

Although the checkpoints that allow transition from one cell cycle phase to another are determined in the same way for any cell, transition times fluctuate, particularly in malignant cells. This is particularly evident in G_1 phase, which accounts for most of the variability of the cell cycle time (and in G_0, whenever it exists). Therefore, instead of using deterministic variables (e.g., subpopulation volume or number of cells in the subpopulation) to describe compartments (see also [42]), in models introduced in this chapter mean values are used, underscoring stochastic nature of model parameters.

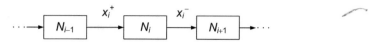

Fig. 2.2 General idea of compartmental modeling. Each compartment represents a subpopulation; cells move from one subpopulation to another, constituting outflow for a preceding compartment and inflow into the subsequent compartment

Let us denote by $N_i(t)$ the average number of cells in the i-th compartment at time t, and by $x_i^+(t)$ and $x_i^-(t)$, the average flow rates of cells into and out of this compartment, respectively (Fig. 2.2). We will also assume that cell death due to apoptosis or other reasons is neglected (this will be modified when drug action is taken into account) and new cells appear only as a result of cell division. Then, the change dN_i of the mean population size in time period $(t, t + dt)$ is given by

$$dN_i = x_i^+ dt - x_i^- dt, \tag{2.1}$$

which can be rewritten as

$$\dot{N}_i(t) = \frac{dN_i}{dt} = x_i^+ - x_i^- \tag{2.2}$$

The simplest models arise if the transit times through each compartment are exponentially distributed. Then, if a_i is the parameter of the exponential distribution, equal to the inverse of the average transit time through the i-th compartment,

$$x_i^-(t) = a_i N_i(t). \tag{2.3}$$

If the preceding compartment is numbered $i - 1$, then $x_{i-1}^- = x_i^+$ and

$$\dot{N}_i(t) = -a_i N_i(t) + a_{i-1} N_{i-1}(t). \tag{2.4}$$

for $i = 2, 3, \ldots, n$ where n is the number of compartments [111]. If the compartmental model is to represent cell cycle, we must take into account that after mitosis newly born cells enter the G_1 phase. Therefore, the boundary condition for the obtained set of equations is given by

$$\dot{N}_1(t) = -a_1 N_1(t) + 2a_n N_n(t) \tag{2.5}$$

If the quiescent G_0 phase is neglected, then the flowchart representing such model takes the shape presented in Fig. 2.3.

If the G_0 phase is taken into account, not all compartments are in a sequential order. For example, if separate compartments represent the cells in G_0, G_1, S, and G_2/M phases, under the exponentiality assumption, the number of cells in various

Fig. 2.3 Flowchart
representing a general model
of cell cycle without the G_0
phase

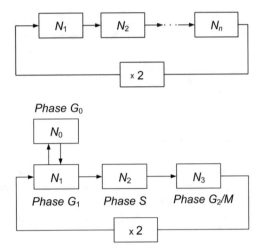

Fig. 2.4 Example of a
four-compartmental model
without the control action

cell cycle compartments versus time, in the absence of external stimuli, is expressed
by a system of the following ordinary linear differential equations (see also Fig. 2.4):

$$\begin{cases} \dot{N}_0 = a_{10}N_1 - a_{01}N_0 \\ \dot{N}_1 = 2a_{31}N_3 - a_{10}N_1 - a_{12}N_1 \\ \dot{N}_2 = a_{12}N_1 - a_{23}N_2 \\ \dot{N}_3 = a_{23}N_2 - a_{31}N_3 \end{cases} \qquad (2.6)$$

2.2   Optimization of Cell Cycle Dependent Chemotherapy

In Chap. 1, a general idea of finding a treatment protocol as a solution to a control
optimization problem was introduced. To do that, we must incorporate treatment
in the form of control variables into models describing cancer growth. There is no
universal way to do that, and the way these variables are introduced into models
depends, among others, on what is the actual goal of exerting a particular control
action. Therefore, in the sections that follow, first the general mathematical terms
representing control actions are presented and examples of their application are
shown.

2.2.1 Compartmental Models of Cell Cycle with Control Action

The substances that are used in chemotherapy can be divided into several groups, depending on the mode of action (Fig. 2.5):

1. Killing agents, represented by u, include the G_2/M specific agents, such as the so-called spindle poisons vincristine, vinblastine or bleomycin, which destroy the mitotic spindle [23] and paclitaxel [43] or 5-fluorouracil [24] also affecting cells mainly in division. Killing agents also include S-specific drugs such as cyclophosphamide [43] or methotrexate [28] acting mainly in the DNA replication phase, cytarabine, rapidly killing cells in phase S by inhibition of DNA polymerase via competition with deoxycytosine triphosphate [34].
2. Blocking drugs, represented by v, including, among others, antibiotics doxorubicin, daunomycin, and idarubicin, which cause the G_1/S block by interfering with formation of the polymerase complex or by hindering the separation of the two polynucleotide strands in the double helix [6]. Another blocking agent is hydroxyurea [37] which synchronizes cells by causing inhibition of DNA synthesis in the phase S and holding cells in G_1.

Fig. 2.5 Possible control actions in the cell cycle

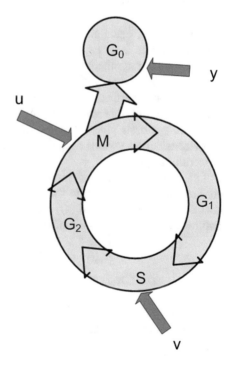

3. Recruitment agents, represented by y, whose role is to recruit cells into a
   particular cell cycle phase, before a killing agent is administered. It is especially
   important for cancer cells in the quiescent G_0 phase, which are resistant to
   cytotoxic agents. Such action was demonstrated [7] for granulocyte colony
   stimulating factors (G-CSF), granulocyte macrophage colony stimulating factors
   (GM-CSF), and interleukin-3 (IL-3), especially when combined with the Human
   Cloned Stem Cell Factor (SCF).

Some drugs have multimodal action. For example, although mostly active in
specific phases, cyclophosphamide and 5-fluorouracil kill or block cells also in other
phases. This makes possible to combine them with other cycle-specific agents [19].
On the other hand, some antimitotic agents such as curacin A [70] act by increasing
the S phase transition and decreasing the M phase transition, which results in a G_2
block.

Control of the cell cycle via the cytotoxic and cytostatic agents is introduced
into the system of linear ordinary differential equations (ODEs) as a multiplicative
term perturbing the exit and/or entrance rates from/to a cell cycle phase [111, 128].
It follows from the fractional kill hypothesis (Skipper hypothesis) which states that
a defined chemotherapy concentration, applied for a defined time period, kills a
constant fraction of the cells in a population, independent of the absolute number of
cells [25, 102].

The resulting equations include linear combinations of products of state and
control variables (see, e.g., [88]). Similar models in discrete-time were used for
optimization of radiotherapy protocols [12, 62].

Usually, the G_2 and M phases are combined in the models into one compartment.
In the two-compartment model, G_0, G_1, and S form the other compartment,
while a range of three-compartment models arise by separately treating either
the DNA synthesis phase S or the dormant phase G_0. These alternative models
allow considering various drugs used in chemotherapy such as killing, blocking,
or recruiting agents.

In a compartmental model, the application of a killing agent acting on the i-th
compartment is equivalent to the death of a fraction of cells in the flow between
compartments. Therefore, if $u(t)$ denotes drug action, only a fraction $1 - u(t)$ of the
outflow from the i-th compartment contains live cells ($0 \leq u(t) \leq 1$):

$$\dot{N}_i(t) = -a_i N_i(t) + a_{i-1} N_{i-1}(t), \tag{2.7}$$

$$\dot{N}_{i+1}(t) = -a_{i+1} N_{i+1}(t) + (1 - u(t)) a_i N_i(t), \tag{2.8}$$

where $u(t) = 0$ and $u(t) = 1$ represent no action and maximum drug effect (related
to maximum feasible drug dose), respectively.

Of course, it is a simplified model, as in real life the cell affected by the drug may survive but be unable to proliferate. The fact that it continues its cell cycle for some time makes estimation of model parameters even more difficult.

Incorporation of the effects of blocking agents consists in reduction of flows between compartments. The flow between i-th and $i + 1$-th compartment is reduced to a fraction $v(t)$ of a flow taking place when no drug is used. The remaining fraction is blocked in the i-th compartment:

$$\dot{N}_i(t) = -(1 - v(t))a_i N_i(t) + a_{i-1} N_{i-1}(t), \tag{2.9}$$

$$\dot{N}_{i+1}(t) = -a_{i+1} N_{i+1}(t) + (1 - v(t))a_i N_i(t). \tag{2.10}$$

It might be difficult to completely block the flow between compartments and, as a result, $0 \leq v \leq v_m < 1$. There are two reasons for that. First, when considering the population of cells, a complete inhibition would mean that each cell in the population has been blocked. This, in turn, would require interaction of drug molecules with each cell, which is stochastic and one cannot assume that its probability is one. Second, if the interaction occurred, some cells might escape the block due to the stochastic nature of intracellular processes that govern the cell cycle checkpoints. Only if the drug concentration is very high (thus making the probability of affecting each cell close to 1) and the drug is very efficient, the value of v_m is close to 1.

The action of a recruitment agent, leading to alteration of the transit time $y(t)$ can be modeled by a multiplicative change of a parameter corresponding to the average time of transit through a given compartment. If $y(t) > 1$, then the transit time has been shortened, if $y(t) < 1$, it has been increased. The former case corresponds to a recruitment action. For example, in the case of recruitment of cancer cells from the quiescent phase G_0, represented by the i-th compartment, the resulting set of equations takes the following form:

$$\dot{N}_i(t) = -y(t)a_i N_i(t) + a_{i-1} N_{i-1}(t), \tag{2.11}$$

$$\dot{N}_{i+1}(t) = -a_{i+1} N_{i+1}(t) + y(t)a_i N_i(t), \tag{2.12}$$

where $y(t) > 1$. The case when $y(t) < 1$, on the other hand, describes cell arrest in the i-th compartment and because of the exponentiality assumption, it is mathematically identical with equations modeling a blocking action. Then, in order to keep a consistent interpretation for all control variables, where 0 denotes no drug being used, it is convenient to use another variable, $w = y - 1$, $0 \leq w \leq w_{max}$. Therefore, in subsequent models taking into account cell recruitment, the following forms of (2.11) and (2.12) are used:

$$\dot{N}_i(t) = -(w(t) + 1)a_i N_i(t) + a_{i-1} N_{i-1}(t), \tag{2.13}$$

$$\dot{N}_{i+1}(t) = -a_{i+1} N_{i+1}(t) + (w(t) + 1)a_i N_i(t). \tag{2.14}$$

The important property which is satisfied in all models in this chapter is that the control systems, which they represent, preserve positivity [50, 58, 59], i.e. for any admissible control and positive initial states, the state remains positive for all times $t \geq 0$. A sufficient condition for this property to hold (for example, see [58]) is satisfied for any compartmental model whose dynamics is given by balance equations where the diagonal entries correspond to the outflows from the i-th compartments (are negative) and the off-diagonal entries represent the inflows from the i-th into the j-th compartment, $i \neq j$ (are nonnegative).

2.2.2   Models with a Single Phase-Specific Killing Agent

The simplest models that take into account phase specificity of the drugs are two-compartmental. One of the compartments is chosen to contain cells in the phase targeted by the drug while the other one, cells in all other phases. For example, if the killing agents are assumed to affect the G_2/M phase (e.g., taxoids or vinca alkaloids) the model is described by the following set of equations [64, 110, 114]:

$$\begin{cases} \dot{N}_1(t) = -a_1 N_1(t) + 2(1 - u(t))a_2 N_2(t), \ N_1(0) = N_{10} > 0, \\ \dot{N}_2(t) = -a_2 N_2(t) + a_1 N_1(t), \qquad\qquad N_2(0) = N_{20} > 0, \end{cases} \quad (2.15)$$

where N_1 is the combined average number of cells in the G_1 and S phases, while N_2 in G_2 and M phases.

If the drug used affects cells in S phase (like topoisomerase inhibitors [126, p. 138]), a separate compartment has to be created for cells in the S phase. Then the model takes the following form:

$$\begin{cases} \dot{N}_1(t) = -a_1 N_1(t) + 2a_3 N_3(t), \qquad\qquad N_1(0) = N_{10} > 0, \\ \dot{N}_2(t) = -(1 - v(t))a_2 N_2(t) + a_1 N_1(t), \ N_2(0) = N_{20} > 0, \qquad (2.16) \\ \dot{N}_3(t) = -a_3 N_3(t) + (1 - v(t))a_2 N_2(t), \ N_3(0) = N_{30} > 0, \end{cases}$$

where N_1, N_2, and N_3 are the average numbers of cells in the G_1, S, and G_2/M phases, respectively and $v(t)$ denotes blocking drug action.

2.2.3 Multiple Drug Therapy and Control

While the killing agent was the only control considered in the two-compartment model, in three-compartment models, also a blocking agent which slows down the transit through S and then releases cells when another G_2M specific anticancer drug has a maximum killing potential (the so-called synchronization [21]) may be considered. This strategy may have the additional advantage of protecting the normal cells which would be less exposed to the second agent, due to less dispersion and faster transit through G_2M [2, 37]. Then, the model includes separate compartments for the G_1, S, and G_2M phases, represented by N_1, N_2, and N_3 variables, respectively [64, 110, 114]:

$$
\begin{cases}
\dot{N}_1(t) = -a_1 N_1(t) + 2(1 - u(t))a_3 N_3(t), \ N_1(0) = N_{10} > 0, \\[2mm]
\dot{N}_2(t) = -(1 - v(t))a_2 N_2(t) + a_1 N_1(t), \ \ N_2(0) = N_{20} > 0, \\[2mm]
\dot{N}_3(t) = -a_3 N_3(t) + (1 - v(t))a_2 N_2(t), \ \ N_3(0) = N_{30} > 0,
\end{cases}
\qquad (2.17)
$$

where $u(t) \in [0, 1]$ and $v(t) \in [0, v_m]$, $v_m \leq 1$, represent killing and blocking (synchronizing) drugs, respectively. Dynamics and control of the model (2.15)–(2.17), as well as other possible models including recruitment of cells from G_0 to G_1, were analyzed in [64, 111, 113, 115, 121].

The drawback of models (2.15), (2.16), or (2.17) is that none of them takes into account existence of a quiescent subpopulation of cells in the G_0 phase. This remains a serious issue in chemotherapy of leukemias, where leukemic stem cells remain quiescent and are not sensitive to most cytotoxic agents [24, 68, 85]. Similar findings were reported for breast and ovarian cancers, e.g., in [28, 43]. As indicated by these authors, the insensitivity of dormant cells to the majority of anticancer drugs and high percentage of tumor cells at rest is a fact which, if ignored, leads to clinical problems and inaccurate theoretical considerations. Experiments with cytarabine [34] indicated that while injected twice during cell cycle or combined with doxorubicin or anthracyclines, cytarabine led to serious reduction of leukemic burden without an evident increase of negative effect on normal tissues. This gain was attributed to the recruitment of leukemic cells from G_0 by Ara-C. It also became possible to efficiently recruit quiescent cells into the cycle using cytokines [123].

Cytokines are substances playing a role in the regulation of normal hemopoiesis, such as G-CSF, GM-CSF, and especially IL-3 combined with SCF. Then, a cytotoxic agent such as cytarabine or anthracyclines may be used. The three-compartment model that describes such therapy uses separate compartments for the G_0, G_1, and $S + G_2M$ phases and includes such a recruiting agent. Moreover, it also enables analysis of the change of the transit time through the G_0 phase due to the feedback mechanism that recruits the cells into the cycle when chemotherapy is applied.

This type of recruitment can be incorporated into a model with different compartments:

$$
\begin{cases}
\dot{N}_0(t) = -(1 + w(t))a_0 N_0(t) + 2b_0(1 - u(t))a_2 N_2(t), \\
\quad N_0(0) = N_{00} > 0, \\[2mm]
\dot{N}_1(t) = -a_1 N_1(t) + (1 + w(t))a_0 N_0(t) + 2b_1(1 - u(t))a_2 N_2(t), \\
\quad N_1(0) = N_{10} > 0, \\[2mm]
\dot{N}_2(t) = -a_2 N_2(t) + a_1 N_1(t), \\
\quad N_2(0) = N_{20} > 0,
\end{cases}
\tag{2.18}
$$

where N_0, N_1, N_2 are the average numbers of cells in G_0, G_1 and $S + G_2 M$ compartments, b_0 and b_1 are the probabilities of the daughter cell entering G_0 and G_1 after division, respectively.

In a similar way we may model other manipulations of the cell cycle, such as the use of triterpenoids to inhibit proliferation and induce differentiation and apoptosis in leukemic cells [69].

Finally, if all three types of drug actions (killing, blocking, and recruitment) are to be included in the description of the dynamics of cancer population, the simplest model is described by the following equations:

$$
\begin{cases}
\dot{N}_0(t) = -(1 + w(t))a_0 N_0(t) + 2b_0(1 - u(t))a_3 N_3(t), \\
\quad N_0(0) = N_{00} > 0, \\[2mm]
\dot{N}_1(t) = -a_1 N_1(t) + (1 + w(t))a_0 N_0(t) + 2b_1(1 - u(t))a_3 N_3(t), \\
\quad N_1(0) = N_{10} > 0, \\[2mm]
\dot{N}_2(t) = -(1 - v(t))a_2 N_2(t) + a_1 N_1(t), \\
\quad N_2(0) = N_{20} > 0, \\
\dot{N}_3(t) = -a_3 N_3(t) + (1 - v(t))N_2(t),
\end{cases}
\tag{2.19}
$$

where N_0, N_1, N_2, and N_3 are the average numbers of cells in G_0, G_1, S, and $G_2 M$ compartments, respectively.

2.2.4   Optimization of Treatment Protocols

The general goal of therapy, as stated in Chap. 1, is maximization of tumor cure probability. The corresponding performance index in the form of (1.2) is equal to the number of cells at the end of therapy. Taking into account that in most models, as shown in the preceding sections, cancer cells are classified into different types, the index to be minimized may be defined by

$$J = \sum_i r_i N_i(T_k),$$ (2.20)

where $N_i(T_k)$ denote the number of cells of i-th type (in the i-th compartment), r_i are weight factors associated with them, and T_k denotes the fixed therapy length. Flexibility in choice of different r_i's could be used to control the fate of cells in selected compartments. For example, in the case of recruitment, by choosing $r_0 > 0$ and $r_i = 0$ for $i \neq 0$ we favor protocols directed at eradication of quiescent cells.

However, one has to take into account that the cumulative dose of drug (or radiation) is bounded, i.e.

$$\int_0^{T_k} u \, dt \leq \Xi.$$ (2.21)

Due to the particular form of the constraint (2.21) [82] (see also Appendix B), the performance index to be minimized takes the following form:

$$J = \sum_i r_i N_i(T_k) + r_u \int_0^{T_k} u(t) dt,$$ (2.22)

where r_u is a Lagrange multiplier related to constraint Ξ.

In case of multiple drug therapy or combined therapy (see Sect. 3.4), the goal is to minimize

$$J = \sum_i r_i N_i(T_k) + \sum_{k=0}^m r_{uk} \int_0^{T_k} u_k(t) dt,$$ (2.23)

where $r_{uk} \geq 0, k = 1, 2, \ldots m$ are weighing factors associated with k-th therapeutic action. All control variables are bounded:

$$0 \leq u_k(t) \leq u_{k_{\max}}, \qquad k = 1, \ldots, m$$ (2.24)

Expression (2.22) is a special case of (2.23).

Such form of the performance index allows an intuitive interpretation of the optimization goal of minimizing the resistant cancer subpopulation at the end of therapy with simultaneous minimization of the cumulative negative effect of the drug represented by the integral component. Another optimization problem arises when the so-called containment treatment is considered, in which the goal is to maintain the number of tumor cells at the lowest possible level while preserving an acceptable number of normal cells [15]. This leads to periodical protocols to be discussed later on in this chapter.

While some authors solve the resulting optimization problem assuming a free terminal time T_k (e.g., [78]), it seems more reasonable to assume a fixed, finite T_k. The most important reason is that otherwise the optimized T_k might be too large to be biologically applicable.

Yet another approach to state the optimization problem for finding therapy protocols involves introduction of the quadratic performance index, in the form of (e.g., [3])

$$J = \int_0^{T_k} \left[u^2(t) + \sum_i N_i(t) \right] dt. \tag{2.25}$$

In that way some authors aim at finding an optimal closed-loop control. Such approach, however, does not seem justified. First, the quadratic term used in the performance index represents either energy costs or variance from a given value. It has no direct biological interpretation. Moreover, continuous measurement of the system state is not feasible, and will not be feasible in a foreseeable future. That renders application of a continuous closed-loop control impossible. At best, such system could be designed as a sampled data system. Then, if sampling period was realistic from a clinical point of view, the results would be very interesting and worth discussing. Taking into account, however, that "realistic" in this case would mean taking only a few measurements during the therapy, the idea of closed-loop control seems out of the question, at least for now.

Necessary conditions of optimality can be found using Pontryagin maximum principle [97] (see Appendix B for more details). To explain how it should be applied, let us focus on the simplest phase-specific model given by (2.15), for which the aim is to minimize the performance index (2.22).

Hamiltonian in this case is given by

$$\begin{aligned} H &= p_1(-a_1 N_1 + 2(1-u)a_2 N_2) + p_2(-a_2 N_2 + a_1 N_1) + r_u u \\ &= u\left(r_u - 2a_2 p_1 N_2\right) - p_1 a_1 N_1 + p_2(-a_2 N_2 + a_1 N_1) \end{aligned} \tag{2.26}$$

and it is linear with respect to the control variable $0 \le u \le 1$. Therefore, the necessary condition for the optimal control is

$$u(t) = \begin{cases} 0 & \text{if } f(t) > 0, \\ 1 & \text{if } f(t) < 0, \\ \text{singular} & \text{if } f(t) = 0, \end{cases} \tag{2.27}$$

where $p = [p_1, p_2]^T$ is the co-state vector and $f(t)$ denotes the so-called switching function, which in the case of the Hamiltonian (2.26) takes the form

$$f(t) = r_u - 2a_2 p_1 N_2. \tag{2.28}$$

The co-state variables are defined by the conjugate equations

$$\dot{p}_1(t) = a_1(p_1(t) - p_2(t)), \qquad\qquad p_1(T) = r_1,$$
$$\dot{p}_2(t) = a_2(p_2(t) - 2p_1(t)(1 - u(t))), \; p_2(T) = r_2, \tag{2.29}$$

The structure of the optimal controls is determined by the switching function and its derivatives. For instance, if $f(t) = 0$, but $\dot{f}(t) \neq 0$, then the control has a switch at time t. If $f(t) = 0$ and $\dot{f}(t) = 0$ over some open interval of t, $I = (\tau, \tau + \delta\tau)$, $\delta\tau \neq 0$, then the so-called *singular* control constitutes the solution in this interval. Therefore, in order to analyze the structure of the optimal controls, we need to analyze the switching function and its derivatives. Singular controls can be calculated by differentiating the switching function with respect to time until the control variable explicitly appears in the derivative. If the order of this derivative is denoted by $l = 2k$ (it has been proved that l is even [71]; see also Appendix B), then k is called the order of the singular arc on the interval I.

For example, the first derivative of the switching function (2.28) is given by

$$\dot{f} = -2a_2\left(\dot{p}_1 N_2 + p_1 \dot{N}_2\right). \tag{2.30}$$

Substituting (2.29) and (2.15) for $\dot{p}_1$ and $\dot{N}_2$, respectively, yields

$$\dot{f} = -2a_2\left(a_1(p_1 - p_2)N_2 + p_1(-a_2N_2 + a_1N_1)\right). \tag{2.31}$$

Since this derivative does not contain control u, second derivative must be calculated:

$$\ddot{f} = -2a_2\left[a_1(\dot{p}_1 - \dot{p}_2)N_2 + a_1(p_1 - p_2)\dot{N}_2 \right.$$
$$\left. + \dot{p}_1(-a_2N_2 + a_1N_1) + p_1(-a_2\dot{N}_2 + a_1\dot{N}_1)\right] \tag{2.32}$$

Since $\dot{N}_1$ in (2.15) explicitly depends on u, the control variable will appear in $\ddot{f}$ and $l = 2$. Therefore, singular controls are of the order $k = 1$. The necessary condition for optimality of a singular arc of order k is given by the generalized Legendre–Clebsch condition [71] (see also Appendix B):

$$(-1)^k \frac{\partial}{\partial u} \frac{d^{2k}}{dt^{2k}} \frac{\partial H}{\partial u} \geq 0. \tag{2.33}$$

In the case considered above, $\frac{\partial H}{\partial u} = f(t)$, which yields

$$\frac{\partial}{\partial u} \frac{d^2}{dt^2} \frac{\partial H}{\partial u} = 4a_1 a_2 > 0, \tag{2.34}$$

so the condition (2.33) is not satisfied and therefore singular arcs are not optimal. Consequently, finding the optimal solution requires finding a numerical solution of the two point boundary value problem (TPBVP) which may be performed using Mohler's switching-time-variation method (STVM) [88, 101], semianalytical shooting algorithm [114], or gradient-type methods [40, 41, 84]. Other possible algorithms are based on discretization of the optimal control problem, solving a large scale nonlinear programming problem resulting from this, and subsequent optimization of switching times (e.g., [46, 47]). It was shown [118] that in the model discussed above, the possible number of switches is 0, 1, 2 and infinity (the latter actually representing a periodic solution) and in each case it is possible to find numerically a solution yielding similar values of the performance index. Moreover, under the assumption of one switch, two solutions are optimal. Then, choice of the final solution can be based on additional constraints on the process of reducing the tumor burden. One of them can be the requirement that the population of cancer cells decreases faster than at a given rate. This leads to another class of optimization problems, in which a periodic solution is sought for. This is particularly important from a clinical point of view, where periodic drug administration seems to be a standard procedure. Such solutions usually yield a larger value of the performance index, constituting a worse solution than the optimal one, at least mathematically. However, if the difference between the suboptimal periodic and optimal solutions, in terms of the performance index is not large, the former usually are easier to be applied in clinical practice. Moreover, if it provides smaller tumor size than at the beginning of treatment, it usually is a better solution with respect to other criteria that have not been included in the original optimization problem formulation. Sample solutions are shown in Fig 2.6.

Complete analysis of solutions satisfying the necessary conditions for optimal control in such model was performed in [116]. All possible solutions, resulting from various values of initial conditions $[N_{10}, N_{20}]$ and weighing factors in the performance index r_1, r_2 are presented in Fig. 2.7, where $\Phi_N = \arctan \frac{N_2}{N_1}$ and $\Phi_p = \arctan \frac{p_2}{p_1}$ denote polar coordinates used to reduce the model dimension. Rigorous mathematical analysis of optimality conditions by means of the field of neighboring extremals led to the proof that only protocols with maximum one switch are optimal [74, 75].

Taking into account the particular form of the performance index to be minimized, given by either (2.22) or (2.23), which are linear with respect to the control variables, and the form of the model equations, which is bilinear with respect to state and control variables, it is clear that the control variables in any model in this section satisfy the condition (2.27). For example, when the model (2.17) of cell arrest in S and killing in G_2/M phases is considered, with performance index given by (2.23), the so-called *bang-bang* solution (in which the control switches between its maximum and minimum values provided by constraints) found based on the maximum principle has the following form

$$u(t) = \begin{cases} 0 \text{ if } r_u - 2a_3 p_1 N_3 > 0, \\ 1 \text{ if } r_u - 2a_3 p_1 N_3 < 0, \end{cases} \tag{2.35}$$

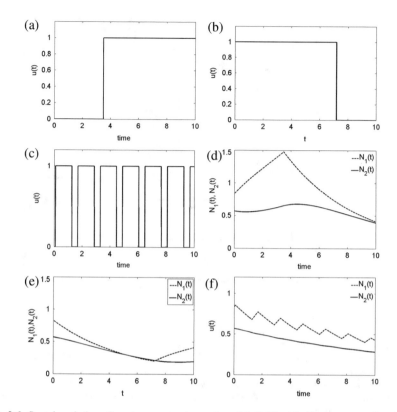

Fig. 2.6 Sample solutions for a two-compartmental model (2.15): (**a**)–(**c**) represent control $u(t)$ that satisfies the necessary conditions for optimality and (**d**)–(**f**) illustrate corresponding sizes of cell populations. Parameter values come from [41]

$$v(t) = \begin{cases} 0 & \text{if } r_v + p_2 - p_3 > 0, \\ v_{\max} & \text{if } r_v + p_2 - p_3 < 0, \end{cases} \qquad (2.36)$$

where the r_u and r_v are weighing factors associated with controls u and v, respectively, and the co-state vector satisfies the following set of equations,

$$\begin{aligned}
\dot{p}_1(t) &= a_1(p_1(t) - p_2(t)), & p_1(T) &= r_1, \\
\dot{p}_2(t) &= a_2(p_2(t) - p_3(t))(1 - v(t)), & p_2(T) &= r_2, \\
\dot{p}_3(t) &= a_3(p_3(t) - 2p_1(t))(1 - u(t)), & p_3(T) &= r_3,
\end{aligned} \qquad (2.37)$$

Once again, before attempting to find optimal bang-bang control, feasibility of singular solutions has to be checked. In this case, the generalized Legendre–Clebsch condition (2.33) still applies to the first control u, if we freeze the second control v. Assuming v is constant, it can be shown that a singular control u must be of order 2,

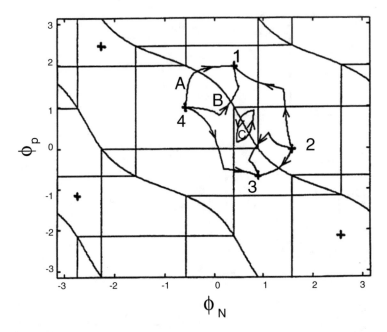

Fig. 2.7 All possible solutions to the TPBVP arising in optimization problem for model (2.15) [116]. There are four equilibrium points: two stable points 1 and 3 and two unstable points 2 and 4. Depending on the initial condition, the solution can contain either (1) no switches (e.g., part of the trajectory from 4 to 1 denoted by *A*); (2) one switch (e.g., part of the trajectory from 2 to 1); (3) two switches (e.g., part of the trajectory from 4 to 1 denoted by *B*); or (4) periodic switches (in the region denoted by *C*)

but again (2.33) is violated. Direct, but longer calculations [74] verify that

$$\frac{\partial}{\partial u} \frac{d^4}{dt^4} \frac{\partial H}{\partial u} = -12a_1 a_2 a_3^2 (1 - v)(a_1 + a_2(1 - v))p_1(t)N_2(t) < 0. \tag{2.38}$$

Furthermore, if the control v is singular on an interval I, then it is clear that u also must be singular on I. In this case we use a necessary condition for optimality, the so-called Goh condition [71] that on I we have

$$\frac{\partial}{\partial v} \frac{d}{dt} \frac{\partial H}{\partial u} \equiv 0. \tag{2.39}$$

However, a direct calculation gives

$$\frac{\partial}{\partial v} \frac{d}{dt} \frac{\partial H}{\partial u} = 2a_2 a_3 p_1(t)N_2(t) > 0 \tag{2.40}$$

violating the Goh condition [74], so the singular solution can be excluded. Note that these results strongly depend on the fact that states and co-states are positive.

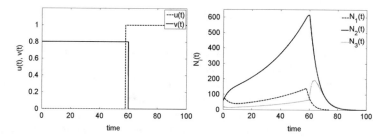

Fig. 2.8 Optimal solutions for an example of a model with blocking and killing drug (**a**) optimal controls, (**b**) corresponding state trajectories. Parameters were taken from [41]

The TPBVP arising may be once more treated numerically [41] by the gradient method in the way similar as for two-compartmental models (see Fig. 2.8).

In the final example, in which the goal is to find optimal control for the model (2.18), the optimal solution satisfies the following:

$$u(t) = \begin{cases} 0 \text{ if } r_u - 2b_0 a_2 N_2 p_0 - 2b_1 a_2 N_2 p_1 > 0, \\ 1 \text{ if } r_u - 2b_0 a_2 N_2 p_0 - 2b_1 a_2 N_2 p_1 < 0, \end{cases} \quad (2.41)$$

$$w(t) = \begin{cases} 0 \quad \text{ if } r_w - a_0 p_0 N_0 + a_0 p_1 N_0 > 0, \\ w_{\max} \text{ if } r_w - a_0 p_0 N_0 + a_0 p_1 N_0 < 0, \end{cases} \quad (2.42)$$

with co-state variables p_0, p_1, and p_2 given by

$$\begin{aligned} \dot{p}_0(t) &= (1 + w(t))a_0(p_0(t) - p_1(t)), & p_1(T) &= r_0, \\ \dot{p}_1(t) &= a_1(p_1(t) - p_2(t)), & p_2(T) &= r_2, \quad (2.43) \\ \dot{p}_2(t) &= a_2 \left[p_2(t) - 2(1 - u(t))(b_0 p_0(t) + b_0 p_0(t)) \right], & p_3(T) &= r_3. \end{aligned}$$

The class of mathematical models based on cell-cycle kinetics, whose examples have been shown above, was introduced in [110, 114] and has been analyzed in numerous papers since (e.g., [64, 111, 113, 115]), from computational and analytical perspective. The general multicompartment model was considered in [74] and once more singular arcs were eliminated using Legendre–Clebsch and Goh conditions.

An interesting finding [118] is that results do not change, at least in qualitative sense, if instead of modeling and minimizing cancer population we rather decide to model and maximize population of cells in critical normal tissues while maximizing the cumulative negative cytotoxic effect.

2.3   Pharmacokinetics and Pharmacodynamics

In the models described above a common simplification is identification of the
drug dose with its concentration in plasma or even with its effects on cancer or
normal cells. In reality these clearly are different values and their mutual relations
are the so-called pharmacokinetics (PK) and pharmacodynamics (PD) [33]. The
first term describes the relationship between the dose and the concentration in the
plasma, while the second, the effect of given concentration on cell viability. Roughly
speaking, PK describes what the cells do to the drug (e.g., transport, metabolization,
removal) and PD illustrates what the drug does to the cells (e.g., DNA damage,
mitotic spindle destruction, myelosuppression). PK and PD are largely influenced
by the diurnal rhythms of proteins and the mRNA levels of which vary according to a
24-h period. They are also disturbed significantly by acquired and genetic resistance
of the cells.

The simplest model of PK was proposed by Bellman [17] in the form of the first
order linear differential equation:

$$\dot{u}^* = -k_{\deg} u^* + k u_{\text{in}} \tag{2.44}$$

where u^* and u_{in} denote the effective (local) and input drug concentrations and $k_{\deg}$
and k denote degradation and effective rates, respectively. It is equivalent to the
assumption that for continuous infusion at a constant rate the growth and decay of
the plasma concentration are exponential with the same time constant $1/k_{\deg}$ [83].

Another model, proposed in [79], is also of the first order, but includes a nonlinear
term:

$$\dot{u}^* = -(k_{\deg} + g u_{\text{in}})u^* + k u_{\text{in}} \tag{2.45}$$

with additional constant parameter $g \geq 0$. This model is also a model of exponential
growth or decay for constant dosages, but it allows for different rates at which the
concentration builds up its maximum level and decays if no drugs are given. The
parameters are still easily related to standard pharmacokinetic data. The maximum
concentration of the drug is given by

$$c_{\max} = k/(k_{\deg} + g), \tag{2.46}$$

which is attained asymptotically for a constant infusion $u \equiv 1$ and the parameters
$k_{\deg}$ and g are related to the times tc_{50} it takes for the concentration to reach 50 %
effectiveness in the following way:

$$tc_{50}^{\text{up}} = \frac{\ln 2}{k_{\deg} + g} \tag{2.47}$$

$$tc_{50}^{\text{down}} = \frac{\ln 2}{k_{\deg}} \tag{2.48}$$

Fig. 2.9 The simplest pharmacodynamics curve. Drug effect is scaled to its maximum value. Drug concentration should be calculated from pharmacokinetics models

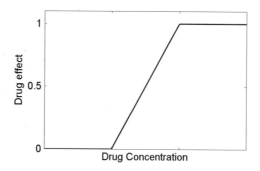

If PK is to be taken into account, in all models described in this section, the control variables u should be replaced by $u^*/u_{in\ max}$ in the model equations, where $u_{in\ max}$ denotes maximum input drug concentration.

It has been proved ([112] for the simplest models, and [76] for the general compartment models) that appending linear PK compartments to the compartmental models of the cell cycle does not change qualitatively the results of optimization. The situation complicates when we take into account different rates in concentration buildup and the drug clearance by the system. One way to overcome the difficulty is by relating the speed of the buildup to the dose of the drug. It leads to a bilinear model of PK which formally is similar to the model of cell cycle kinetics [76] and does not change the structure of the multicompartmental model. The properties of the solution to optimization problem are still unchanged.

The model of pharmacodynamics (PD) describes the effectiveness of the drug. The models previously analyzed identify the effectiveness with the fraction of ineffective cell divisions. Thus it changes between 0 and 1 and PD is modeled by a function linear in some range of concentration and equal to 0, and to 1, respectively, outside the range (Fig. 2.9). For extremes of concentration, more realistic models with saturation include so-called sigmoid E_{max} model, which is a form of a Hill equation [49, 132]:

$$u = \frac{u_{max}u^{*\alpha}}{u_{50}^{*\alpha} + u^{*\alpha}}, \qquad (2.49)$$

where u_{50}^* is the drug concentration for which 50 % of maximum effect is obtained and α is the Hill coefficient of sigmoidicity. Depending on these two parameters values various dose response curves can be obtained (Fig. 2.10).

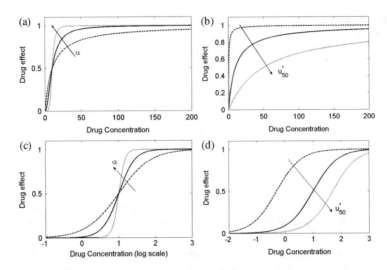

Fig. 2.10 A sigmoid PD model in (**a**), (**b**) linear and (**c**) (**d**) semilog scale. Drug effect is scaled to its maximum value. The *arrows* indicate increasing parameter value. Drug concentration is in arbitrary units

In multidrug treatment such formulations are possible only if the drugs act in different ways. If the drugs act similarly, and at the same time their combined effectiveness depends on the total concentrations and their synergistic properties. Even for the most commonly used drugs these types of interaction are not well recognized and understood. An important finding [76] is that geometric properties of PK/PD models have a direct effect on the qualitative solutions of optimization problems. For linear PK and linear or Michaelis–Menten PD models qualitatively results of optimizations do not change and quantitatively the results do not differ significantly. In the case of sigmoidal PD models nothing changes for convex regions but for the concave ones singular arcs (corresponding to intermediate doses of the drug) could be optimal.

Recently it can be observed an increasing interest in PK/PD modeling related to the mutual dependence between these effects from one side and resistance of cancer cells, especially multidrug resistance, drug–drug interactions including synergy/antagonism relations, and combined therapy from the other [15]. Moreover above-mentioned relationships between PK/PD mechanisms and the circadian clock as well as their influence on the micro-environmental phenomena in cancer development and response to therapies have been studied intensively (e.g., [27, 80, 92, 95, 98]). Incorporating PD mechanisms at the intracellular level is another area of modeling ([26], see also Chap. 5).

2.4   Resonances, Synchronization, Aftereffects

The resonance effect, postulated by several authors (e.g., [2, 38, 130]) as the way
to either maximize the efficacy of treatment or spare the organism's normal cells,
is similar to synchronization using blocking and killing agents, except that the
resonance is usually mentioned in the context of a periodically administered single
agent. At the resonance frequency and dosage, maximum efficiency is achieved.
This tactics may be combined with an attempt to spare the normal cells by taking
advantage of the difference in cell cycle duration of cancer and normal cells. Having
this in mind, let us review several models using the concepts of synchronization
and resonance. A similar effect is also observed in anticancer radiotherapy and is
potentially useful in improving treatment protocols [125].

Models in the form of ODEs ignore the fact that most distributions of transit
times in the cell cycle are not exponential. A more realistic description is using
functional differential or integral equations. A good example is the model used
to solve the simplest control problems arising in chemotherapy, concerning cells
synchronization in normal tissues [37]. The model of the cell cycle assumes
shifted exponential or gamma distributions of the interdivision time. This leads
to a functional differential equation with infinite delay. Such models will be
discussed in Chap. 4 in the context of structured models of cancer growth. Since
the purpose of drug administration is to arrest as many cells as possible in
the specific phase, and not allow them to enter the phase in which they are
exposed to the anticancer drug, the performance index should be maximized. This
problem, which is mathematically more complicated, may be still solved using the
abstract version of the Maximum Principle leading to bang-bang candidates for
optimal chemotherapy strategies. Contrary to the optimization problems discussed
previously, where side effects were considered only in the form of the integral term
in the performance index, in this model the side effects of the drug are central.
Similar reasoning can be applied to model toxicity to the bone marrow, which is
one of the main limiting factors in chemotherapy. Such model was introduced by
Panetta and analyzed as an optimal control problem in [43]. Its authors proposed
an objective quadratic in the control, which led to the protocols with gradually
increasing doses after a no-dose interval, which does not seem to be admissible. The
use of an objective linear in control [79] leads once more to bang-bang strategies,
which remain qualitatively the same even if pharmacokinetics of the drug is taken
into account. Another model, used by Dibrov [38], includes both "probabilistic" and
"deterministic" compartments. Probabilistic compartments are modeled by ODEs,
while the cell density function in the deterministic compartment is a solution
of von Foerster partial differential equation [127]. Solving and substituting its
solution into the equations for the probabilistic compartment leads to a system
of equations with time-delay. Dibrov et al. [38] examined the effect of multiple
periodic treatments and found that the optimal period of drug administration almost
coincides with the mean intermitotic time for normal cells and with this choice
the fraction of normal cells in the sensitive phase at moment of drug injection is

minimized. A deterministic model which represents kinetic heterogeneity of cancer cells belongs to the class of age-structured models [8, 9, 18, 129]. Such models allow a simple representation of cell populations with variable but uncorrelated cell cycle times. They exhibit the asynchronous exponential growth property, which means that population grows asymptotically exponentially with an invariant distribution of the structure variable. More complex models, involving subpopulations with age structure, are required to take into account different cell cycle phases, which enables modeling of the effect of chemotherapy in the age structured context (see, e.g., [73]). Although this class of models has been developed by some authors to find recommendations for chemotherapy planning, it does not lead to general statements since no qualitative methods are known for analysis of such models. Simulation remains the only tool of their investigation. More promise could be related to the approach proposed by Shackney [100] in which randomness occurs in the transitions between different cell maturation stages. Further refinements of this idea enable to find an approximate formula for the distribution of cell cycle time as a function of parameters which may be dependent on chemotherapy schedules [9]. The results are, however, far from being applicable to drawing practical conclusions.

The problem of cytotoxicity reduction by proper scheduling of phase-specific drugs has also been considered in a number of papers by Agur and coworkers (e.g., [2, 2, 30]). Their line of reasoning is based on the so-called Z-method and utilizes a much simpler model in the form of functional equation which may be considered a linear difference equation with variable sampling time. The elimination coefficient Z measuring the treatment efficacy is defined by the ratio of the elimination time of the malignant population and that of critical host population. In [2] it was shown that treatment efficacy is a nonmonotonous function of the cell generation time and the period of drug administration with the maxima occurring when the critical host cell cycle length is a multiple of the chemotherapeutic period. The results imply that short drug-pulses at appropriate intervals may be more efficient than a drug administered at arbitrary intervals or a continuously released drug. Under the condition that the cell cycle parameters of malignant cells have a relatively large variation, the optimal drug protocol might be determined by the host temporal parameters alone and should reduce cytotoxicity even in the case of similar mean cell cycle times of cancer and normal cells. A mathematically elegant model of resonance has been provided by Webb in [130] who used a distributed cell cycle model and the notion of Floquet exponents to demonstrate rigorously the existence of multiple resonances. All these attempts have to be viewed with caution, because of the existence of aftereffects in the action of many cytotoxic agents [65]. Actions of these drugs may extend beyond the span of a single cell cycle. For example, cells blocked in the S-phase of the cell cycle and then released from the block may proceed apparently normally towards mitosis but then fail to divide, or divide but not be able to complete the subsequent round of DNA replication. If such effects are substantial, they are likely to disrupt or complicate the resonances.

2.5 Drug Resistance in Chemotherapy

Drug absorption, distribution, metabolism, and removal are regulated by molecular processes taking place inside all cells. These processes include DNA transcription, RNA processing and transport, RNA translation, the posttranslational modification of proteins, degradation of proteins and intermediate RNA products [5]. Proteins performing the regulatory functions mentioned above are coded by genes. This gives rise to genetic regulatory systems structured by networks of regulatory interactions between DNA, RNA, proteins, and small molecules [36]. A mutation in one or several genes in this regulatory network can significantly alter cell behavior, and may lead to drug resistance. Lists of genes and drugs that may trigger drug resistance can be found in many papers, e.g., in [39, 124].

More specifically, three major mechanisms contributing to drug resistance have been identified [122, 124]:

- decreased uptake of water-soluble drugs which require active transport to enter cells;
- changes in cells that affect the capacity of cytotoxic drugs to kill cells, including alterations in cell cycle, increased repair of DNA damage, reduced apoptosis, and altered metabolism of drugs;
- increased energy-dependent efflux of hydrophobic drugs that can easily enter the cells by diffusion through the plasma membrane;

Emergence of clones of cancer cells resistant to chemotherapy is important for treatment and prevention of systemic spread of disease. Models incorporate drug resistance by dividing cancer cell populations into separate compartments, representing subpopulations characterized by different drug sensitivity. The number of compartments and the rate of flow among them depend on the assumptions about the mechanisms causing drug resistance. Parameters representing growth and death rates may be different in each compartment. In fact, it has been postulated in so-called Norton–Simon hypothesis [90, 91] that clones that gain resistance properties do it at the expense of their growth rate, i.e. their growth rate is negatively correlated with the level of drug resistance.

2.5.1 Simple Models of Drug Resistance

If a mutation leads to complete disruption of a regulatory interaction at any stage, it can give rise to single-step drug resistance, as in [31, 94]. Then, a single additional compartment representing resistant cells should be added to those already included in the model. In the simplest case, one compartment contains drug sensitive cells, while the other—the resistant subpopulations. If N_0 and N_1 denote the average cell numbers in these compartments, the dynamics of the whole system is described as follows (similar equations were presented in [77]):

$$\begin{cases} \dot{N}_0(t) = (1 - 2u(t))\lambda_0 N_0(t) - b N_0(t) + d N_1(t), \ N_0(0) = N_{00} > 0, \\ \dot{N}_1(t) = \lambda_1 N_1(t) + b N_0(t) - d N_1(t), \qquad\qquad N_1(0) = N_{10} > 0. \end{cases} \qquad (2.50)$$

Parameters λ_0 and λ_1 are inverse values of the average cell lifetimes of the sensitive and resistant populations, respectively. Parameters b and d are intensities of mutational events (following Poisson distributions) leading to gain or loss, correspondingly, of the resistance property.

2.5.2   Infinite-Dimensional Models of Drug Resistance

Another model is needed when the so-called gene amplification process is involved in emergence of drug-resistant cells. Additional gene copies are acquired leading to an increase in transcription products and multicompartmental models describing populations with various levels of drug resistance are more appropriate [64, 67, 104, 120]. This approach concerns also cases, when resistance to different drugs in one population is concerned [57, 86]. This latter phenomenon is called multiple drug resistance [39, 72, 81, 93].

Taking into account the biological background presented in this section, a mathematical model combining cancer chemotherapy with drug resistance should be based on the following assumptions:

- the process of developing resistance to an agent is reversible, either by its nature or with the aid of other drugs; irreversible drug resistance is only a special case of a general model;
- if gene amplification causes drug resistance, a wide spectrum of levels of resistance has to be considered (eventually leading to an infinite-dimensional model, discussed further in this chapter).

Another issue in drug resistance is that it can be either intrinsic, exhibited by a subset of cells from the very beginning of a treatment, or acquired, triggered by the chemotherapeutic agent [11]. Both types of resistance can be taken into account in models, where they determine the number and form of equations as well as, in the case of intrinsic resistance, initial conditions. In the former case, this can lead either to classical irreversible mutation models (e.g., [48]) or models of reversible, multistage mechanisms such as gene amplification or other transformations of cancer cell gene expression [10, 63, 66]. Both approaches seem applicable, depending on the biochemical mechanism of resistance and each leads to new mathematical problems. These models are based on the results of experimental works [22, 61, 108, 131].

In mathematical modeling, the emergence of resistance to chemotherapy has been first considered in a point mutation model by Coldman and Goldie in [31, 32], and then in the framework of gene amplification by Harnevo and Agur [52–54]. The main idea was that there existed spontaneous or induced mutations of cancer cells towards drug resistance and that the scheduling of treatment should anticipate these mutations. The consequences of the point mutation model can be translated into simple recommendations, which have been tested in clinical trials.

However, phenomena such as uncontrolled growth of the resistant subpopulation, observed in experiments, could not be fully explained by such models. It was found that cells may gain more than one additional gene copy and the number of these gene copies determines the resistance level of a cell. This has been a basis for another model, introduced below, shown to be a good fit for experimental data where a large number of additional gene copies has been found (e.g., [99]). It proved to be general enough to accommodate different interpretations. Among others, it is a good representation of resistance to methotrexate (MTX)—a clinically important agent being used in the treatment of malignancies including acute lymphocytic leukemia, osteosarcoma, carcinomas of the breast, head and neck, choriocarcinoma, and non-Hodgkin's lymphoma [14]. More recently, gene amplification has been listed as one of the explanations of the so-called copy number variation, observed in many cancers [1].

The model and its properties were discussed in original papers [119, 120]. The compartments are numbered according to increasing numbers of gene copies and corresponding level of drug resistance. Cells of type 0, with no copies of the gene, are sensitive to the cytostatic agent. The resistant subpopulation consists of cells of types $i = 1, 2, \ldots$ with $1, 2, \ldots$ copies of the gene. Due to a mutational event a cell can acquire or lose a copy of a gene that makes it resistant to the agent. Since the number of gene copies per cell is not limited a priori, the number of different cell types is denumerably infinite. This assumption makes it possible to analyze asymptotic properties of the model such as stability, whose conditions otherwise would be subject to an arbitrary threshold in the number of gene copies (see later on in this section). Cell division and the change of the number of gene copies are stochastic processes with the following hypotheses:

- Lifespans (lengths of cell cycle) of all cells are independent exponentially distributed random variables with means $1/\lambda_i$ for cells of type i. Parameter λ_i is assumed to be a constant, independent of drug action. Though chemotherapeutic agents can affect cell lifespans, for models involving average cell numbers, this assumption seems a justifiable simplification.
- Cell of type $i \geq 1$ may mutate in a short time interval $(t, t + dt)$ into a cell of type $i + 1$ with probability $b_i dt + o(dt)$ and into a cell of type $i - 1$ with probability $d_i dt + o(dt)$. Cell of type $i = 0$ may mutate in a short time interval $(t, t + dt)$ into a type 1 cell with probability $\alpha dt + o(dt)$, where α is several orders of magnitude smaller than any of b_i and d_i.

- Drug action results in fraction u_i of ineffective divisions in cells of type i (please note that there is only one drug under consideration here with different effectiveness denoted by subscript i)
- The stochastic process is initiated at time $t = 0$ by a finite population of cells of different types.

These assumptions lead to a stochastic model, describing the dynamics of cell population. If the average number of cells is of interest, the model becomes a system of ODEs:

$$\begin{cases} \dot{N}_0(t) = [1 - 2u_0(t)]\lambda_0 N_0(t) - \alpha N_0(t) + d_1 N_1(t), \\[2mm] \dot{N}_1(t) = [1 - 2u_1(t)]\lambda_1 N_1(t) - (b_1 + d_1)N_1(t) + d_2 N_2(t) + \alpha N_0(t), \\ \quad \cdots, \\[2mm] \dot{N}_i(t) = [1 - 2u_i(t)]\lambda_i N_i(t) - (b_i + d_i)N_i(t) + d_{i+1}N_{i+1}(t) + \\ \qquad\qquad + b_{i-1}N_{i-1}(t), \qquad\qquad\qquad\qquad\qquad\qquad\qquad i \geq 2, \\ \quad \cdots, \end{cases}$$

$$(2.51)$$

where control variables u_i represent drug effect on cells of the type i.

The probability of a mutational event in a sensitive cell is of several orders smaller than the probability of the change in number of gene copies in a resistant cell. Moreover, it has been experimentally found that $b_i < d_i$. These two facts combined might suggest that the variables N_i should converge to zero as $t \to \infty$, if $u_0(t) = 1$. It was the mathematical analysis of an infinite-dimensional model that led to additional conditions of stability for a system, in which all sensitive cells are destroyed by a killing agent (see further in the text).

In [96, 119, 120] only the simplest case has been discussed in which the resistant cells were insensitive to drug's action, and there were no differences among the parameters associated with cells of different types:

$$\begin{cases} \dot{N}_0(t) = [1 - 2u(t)]\lambda N_0(t) - \alpha N_0(t) + d N_1(t), \\[2mm] \dot{N}_1(t) = \lambda N_1(t) - (b + d)N_1(t) + d N_2(t) + \alpha N_0(t), \\ \quad \cdots, \\[2mm] \dot{N}_i(t) = \lambda N_i(t) - (b + d)N_i(t) + d N_{i+1}(t) + b N_{i-1}(t), \\ \quad \cdots, \qquad\qquad\qquad\qquad\qquad\qquad\qquad\qquad\qquad i \geq 2. \end{cases}$$

$$(2.52)$$

The methodology used to analyze this model involved its decomposition into two subsystems: the first one, consisting of only the first equation and the second one, infinite-dimensional, but linear and not depending on control, described by a tridiagonal system matrix. Its particular form allowed the properties of the latter to be fully investigated, which subsequently led to system description by a single integro-differential equation (similar models were also analyzed using semigroup theory, showing possible chaotic behavior [13]). Later, this method was generalized,

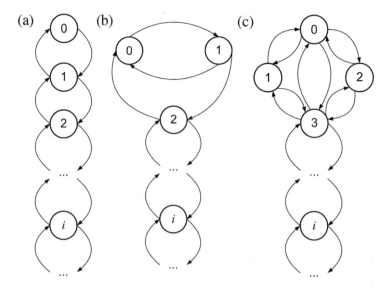

Fig. 2.11 Flows between subpopulations of the model, taking into account (**a**) the original assumptions or partial sensitivity of the resistant subpopulation; (**b**) two-drug chemotherapy; (**c**) phase-specific chemotherapy. In all cases, *numbers* denote cell types

allowing to deal with models taking into account partial drug sensitivities (or resistance), cell cycle phase-specific agents and drug resistance and even multidrug treatment [104, 105, 117]. The graph representing the possible flows between subpopulations is presented in Fig. 2.11. It should be noted that the general model does not accommodate a case, in which two different genes are amplified (this would result in an infinite-dimensional system that is more difficult to analyze). This is justified by the fact that the resistance to each drug depends on different biological processes and only one of them is the increase of the number of gene copies.

The general model is given by the following state equation [117]:

$$\dot{\mathbf{N}} = \left(\mathbf{A} + \sum_{i=0}^{m} u_i \mathbf{B_i} \right) \mathbf{N}. \tag{2.53}$$

where $\mathbf{N} = [N_0 \ N_1 \ N_2 \ \dots \ N_i \ \dots]^T$ is an infinite-dimensional state vector from L^1 space and u_i are control variables. The important assumption is that while the first, finite number of equations may assume any bilinear form, the infinite-dimensional tail is particular, as illustrated in Fig. 2.12. Integer l is the index of the first state variable, for which the corresponding equation takes the form

$$\dot{N}_l = c_1 N_{l-1} + a_2 N_{l_1} + a_3 N_{l+1}. \tag{2.54}$$

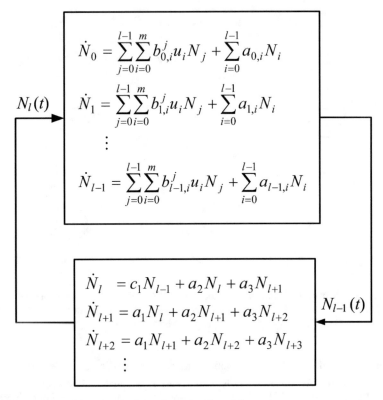

Fig. 2.12 Decomposition of the infinite-dimensional model

Subsequent equations have the same form (Fig. 2.12), matrices $\mathbf{A}$ and $\mathbf{B_i}$ have the following form:

$$\mathbf{A} = \begin{bmatrix} \tilde{\mathbf{A}}_1 & | & \mathbf{0}_1 \\ - & - & - & - \\ \mathbf{0}_2 & | & \tilde{\mathbf{A}}_2 \end{bmatrix}, \qquad (2.55)$$

$$\mathbf{B_i} = \begin{bmatrix} \tilde{\mathbf{B}}_i & | & \mathbf{0}_1 \\ - & - & - \\ & \mathbf{0}_3 & \end{bmatrix}, \qquad (2.56)$$

$$\tilde{\mathbf{A}}_1 = \begin{bmatrix} a_{00} & a_{01} & \dots & a_{0,l-1} & 0 \\ a_{10} & a_{11} & \dots & a_{1,l-1} & 0 \\ \vdots & \vdots & \dots & \vdots & 0 \\ a_{l-1,0} & a_{l-1,1} & \dots & a_{l-1,l-1} & a_{l-1,l} \end{bmatrix},$$

$$\tilde{\mathbf{A}}_2 = \begin{bmatrix} c_1 & a_2 & a_3 & 0 & 0 & \cdots \\ 0 & a_1 & a_2 & a_3 & 0 & 0 & \cdots \\ 0 & 0 & a_1 & a_2 & a_3 & 0 & \cdots \\ \vdots & \vdots & \ddots & \ddots & \ddots & \ddots \end{bmatrix},$$

$$\tilde{\mathbf{B}}_i = \begin{bmatrix} b_{0,0}^i & b_{0,1}^i & \cdots & b_{0,l-1}^i \\ b_{1,0}^i & b_{1,1}^i & \cdots & b_{1,l-1}^i \\ \vdots & \vdots & \cdots & \vdots \\ b_{l-1,0}^i & b_{l-1,1}^i & \cdots & b_{l-1,l-1}^i \end{bmatrix},$$

$\mathbf{u}(t)$ is the m-dimensional control vector $\mathbf{u} = [u_0 \ \ u_1 \ \ u_2 \ \ \cdots \ \ u_{m-1}]^T$, $\mathbf{0}_1$, and $\mathbf{0}_2, \mathbf{0}_3$ are zero matrices of dimensions $l \times \infty$, $\infty \times l - 1$, and $\infty \times \infty$, respectively ($l > m$). Such system decomposition is convenient for analysis and optimal control synthesis.

First, let us consider the infinite-dimensional tail without the influx of cells N_{l-1}:

$$\begin{cases} \dot{N}_l(t) & = a_2 N_l(t) + a_3 N_{l+1}(t), \\[2mm] \dot{N}_{l+1}(t) & = a_1 N_l(t) + a_2 N_{l+1}(t) + a_3 N_{l+2}(t), \\[2mm] & \cdots, \\[2mm] \dot{N}_i(t) & = a_1 N_{i-1}(t) + a_2 N_i(t) + a_3 N_{i+1}(t), \\[2mm] & \cdots . \end{cases} \tag{2.57}$$

It is important to note that the model parameters satisfy the following relations: $a_3 > a_1 > 0$ and $a_2 < 0$, which is always true in the biological application that is considered here. However, a complete problem analysis is possible when no additional conditions are imposed on parameters a_1, a_3, using the same line of reasoning.

Using methods similar to those in [107, 120], it is possible to show that for initial condition $N_i(0) = \delta_{ik}$ (Kronecker delta)—that is, $N_k(0) = 1$, $N_i(0) = 0$ for $i \neq k$ ($k \geq l$ is an arbitrarily chosen number)—the following relations hold true:

$$N_l^k(s) = \frac{1}{a_3} \left(\frac{s - a_2 - \sqrt{(s-a_2)^2 - 4a_1 a_3}}{2a_1} \right)^{k-l+1}, \tag{2.58}$$

$$N_\Sigma^k(s) = \frac{1}{s - (a_1 + a_2 + a_3)} \left[1 - \left(\frac{s - a_2 - \sqrt{(s-a_2)^2 - 4a_1 a_3}}{2a_1} \right)^{k-l+1} \right], \tag{2.59}$$

where $N_l^k(s)$ and $N_\Sigma^k(s)$ are Laplace transforms of $N_l^k(t)$ and $\sum_{i \geq l} N_i^k(t) = N_\Sigma^k(t)$, respectively. In both cases, the superscript k is introduced to indicate the index of the state variable with nonzero initial condition. Now, let us assume that $k = l$. After calculating the inverse Laplace transform the following formulae are obtained:

$$N_l^l(t) = \frac{1}{a_3} \left(\sqrt{\frac{a_3}{a_1}} \right) \frac{I_1 \left(2\sqrt{a_1 a_3} t \right)}{t} \cdot e^{a_2 t}, \tag{2.60}$$

$$N_\Sigma^l(t) = \sum_{i \geq l} N_i^l(t)$$
$$= e^{[(a_1 + a_2 + a_3)t]} \cdot \left[1 - \left(\sqrt{\frac{a_3}{a_1}} \right) \int_0^t \frac{I_1 \left(2\sqrt{a_1 a_3} \tau \right)}{\tau} e^{[-(a_1 + a_3)\tau]} d\tau \right],$$

$$\tag{2.61}$$

where $I_1(t)$ is a modified Bessel function of the first order.

It should be emphasized that the assumption about the initial condition does not introduce any additional constraints to the applicability of the model. Due to linearity of the infinite-dimensional tail any finite nonzero initial condition can be incorporated into the final solution, using superposition principle.

Using an asymptotic expansion of (2.61) it has been found [96] that, assuming $a_3 \geq a_1$, a condition for convergence to zero of the autonomous system is given by

$$a_2 \leq -2\sqrt{a_1 a_3}. \tag{2.62}$$

To understand the implications of those conditions, let us rewrite them using biological parameter notation. They can be presented as

$$d \geq b \tag{2.63}$$

and

$$\sqrt{d} - \sqrt{b} \geq \sqrt{\lambda}, \tag{2.64}$$

In this particular case, if the latter inequality is satisfied, so is the former. The first inequality means that the amplification ratio should not be greater than the deamplification ratio (probability of gaining an additional gene copy is not grater than probability of losing a gene copy). Surprisingly, this is not sufficient for the stability of the autonomous system, since it does not lead to the extinction of the resistant subpopulation when there is no influx of cells from the sensitive compartment. The additional condition implies that the difference of these rates must be large enough so that in most cases the average time needed for the amplification event to happen is larger than the cell lifespan (parameter λ corresponds to the cell lifespan).

Analysis discussed above was performed for a specific initial condition that is nonzero for only one variable, i.e. $N_i(0) = \delta_{i0}$, but it can be expanded on other types of initial conditions. Taking into account linearity of the model, from the superposition principle it stems that the same conditions determine convergence to zero for any initial condition with finite number of nonzero elements, providing stability with finite support. If the drug resistance is caused by the therapy, such analysis is sufficient as it is reasonable to assume zero initial condition for drug resistant cells of any type (and nonzero initial condition for sensitive subpopulation). However, if the drug resistance is intrinsic, there might be drug resistant cells at the beginning of a therapy. Though their number is finite, it is unknown and we can determine the initial condition in the form of a distribution only, possibly leading to infinite number of nonzero initial conditions. Nevertheless, also in that case it is possible to derive stability conditions under assumption that solutions to the equations describing the model are defined in the space of summable sequences [119]. Then, the result is that in addition to the conditions discussed above the initial resistance should also decay, in some sense, with the level of resistance.

Relation (2.58) can be used to determine the following transfer function in the model (2.53):

$$K_1(s) = \frac{N_l(s)}{N_{l-1}(s)} = \frac{c_1}{a_3} \cdot \frac{s - a_2 \sqrt{(s - a_2)^2 - 4a_1 a_3}}{2a_1}. \tag{2.65}$$

Moreover,

$$\sum_{i \geq l} N_i(t) = N_\Sigma^l(t) + N^+(t), \tag{2.66}$$

where

$$N^+(t) = c_1 \int_0^t N_\Sigma^l(t - \tau) N_{l-1}(\tau) d\tau \tag{2.67}$$

and $N_\Sigma^l(t)$ is defined by (2.61).

Let us now introduce the following notation:

$$\hat{\mathbf{B}}_1 = \begin{bmatrix} 0 \\ \vdots \\ 0 \\ a_{l-1,l} \end{bmatrix}, \quad \mathbf{C} = [0, \ldots, 0, 1], \quad \dim \mathbf{C} = l. \tag{2.68}$$

Then, applying control theory techniques [133], for $u(t) = 0$ the following relation holds:

$$\mathbf{K}_2(s) = \frac{N_{l-1}(s)}{N_l(s)} = \mathbf{C}(s\mathbf{I} - \tilde{\mathbf{A}}_1)^{-1} \hat{\mathbf{B}}_1. \tag{2.69}$$

Fig. 2.13 Flowchart of the
system without control

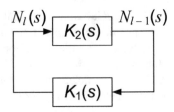

Taking into account the linear form of such a system, it is possible to present the model in the form shown in Fig. 2.13. This makes it possible to analyze the dynamical properties of the closed-loop system.

Let us now consider the problem of stabilization of the system (2.53) by a constant control (which becomes another parameter in such analysis). Then, the transfer function $K_2(s)$ representing the finite dimensional subsystem in Fig. 2.13 takes the form:

$$\mathbf{K}_2(s) = \frac{N_{l-1}(s)}{N_l(s)} = \mathbf{C} \left[s\mathbf{I} - \left(\tilde{\mathbf{A}}_1 + \sum_{i=0}^{m} u_i \tilde{\mathbf{B}}_i \right) \right]^{-1} \hat{\mathbf{B}}_1 \qquad (2.70)$$

The appropriate version of the Nyquist criterion for the infinite-dimensional systems [133] can be applied to find the stability conditions for the model represented by the block diagram shown in Fig. 2.13, in which the constant control u is a parameter. It is stable, if both subsystems, represented by transfer functions $K_1(s)$ and $K_2(s)$, are stable and

$$\sup_{\omega} |K(j\omega)| < 1, \qquad (2.71)$$

where $K(j\omega) = K_1(j\omega) \cdot K_2(j\omega)$.

To show how this condition can be applied, let us consider the model (2.52). The transfer functions $K_1(s)$ and $K_2(s)$ in this case are given by

$$K_1(s) = \frac{\alpha}{d} \cdot \frac{s - (\lambda - b - f) - \sqrt{(\lambda - b - f)^2 - 4bd}}{2b}, \qquad (2.72)$$

$$K_2(s) = \frac{d}{s + 2\lambda u - (\lambda - \alpha)}, \qquad (2.73)$$

The stability condition for the first subsystem is given by (2.64), whereas for the second one by

$$u > \frac{\lambda - \alpha}{2\lambda}. \qquad (2.74)$$

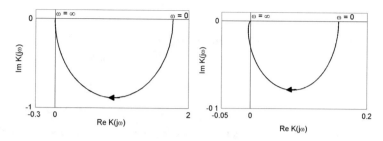

Fig. 2.14 Frequency responses for different values of u: (**a**) $u < u*$ and (**b**) $u > u*$ (*right*)

$|K(j\omega)|$ reaches its supremum for $\omega = 0$ [119]. Therefore, from (2.71) it stems that

$$u* > \frac{1}{2} + \frac{\alpha\left(-\lambda + b + d - \sqrt{((-\lambda + b + d)^2 - 4bd)}\right)}{4b\lambda}, \tag{2.75}$$

or, after algebraic transformation [119]

$$u* > 0.5 + \frac{\alpha}{d - b - \lambda + \sqrt{(b + d - \lambda)^2 - 4bd}} \tag{2.76}$$

Two sample frequency responses, one of which represents an unstable and the other a stable system, are presented in Fig. 2.14.

Stability analysis of the models with constant control, introduced as an additional parameter, provides a feasibility check of a given drug. If the stability conditions require $u > 1$, it means that it is impossible to ensure decay of cancer population. Though an optimization problem can be stated and solved also in that case, applicability of its solution is arguable at best, taking into account that the performance index introduced further in this section is the tumor cure probability.

Recommendation for treatment protocols, regardless of effectiveness of a drug, can be derived only from solving respective optimization problem, as described in the subsequent sections. Another example of an infinite-dimensional model of drug resistance, in the form of a structured population model, is presented in Chap. 4.

2.5.3 Model Transformation and Optimization of Therapy

In order to deal effectively with optimization problems for systems given by (2.53), it is convenient to transform the model description into a set of integro-differential equations [117].

Let us denote

$$\tilde{\mathbf{N}} = \begin{bmatrix} N_0 \\ \vdots \\ N_{l-1} \end{bmatrix} \tag{2.77}$$

and $\mathbf{C_k} = [c_j], c_k = 1, c_j = 0$ for $j \neq k, i = 1, 2, \ldots, l-1$.

Let us also assume the initial conditions $N_i(0) = 0$ for $i > l - 1$ (any finite nonzero initial condition can be incorporated in the final solution). Then, the last equation in the first subsystem, influenced directly by control, as presented in Fig. 2.12, can be transformed into an integro-differential form:

$$\dot{N}_{l-1}(t) = \sum_{j=0}^{l-1} \sum_{i=0}^{m} b_{l-1,i}^j u_i(t) N_j(t) + \sum_{i=0}^{l-1} a_{l-1,i} N_i(t) + a_{l-1,l} \int_0^t k_1(t-\tau) N_{l-1}(\tau) d\tau, \tag{2.78}$$

where $k_1(t)$ is the inverse Laplace transform of $K_1(s)$, given by (2.65).

Other equations can also be rewritten in an analogous way, leading to transformation of the model (2.53) into the following form:

$$\dot{\tilde{\mathbf{N}}} = \mathbf{h}(\mathbf{u}, \tilde{\mathbf{N}}) + \int_0^t \tilde{\mathbf{f}}(\tilde{\mathbf{N}}, t, \tau) d\tau, \quad \tilde{\mathbf{N}}(0) = \tilde{\mathbf{N}}_0, \tag{2.79}$$

where $\mathbf{h}(..)$, $\tilde{\mathbf{f}}(\ldots)$ are the respective l-dimensional vector functions

$$\mathbf{h_k}(\mathbf{u}, \tilde{\mathbf{N}}) = \sum_{j=0}^{l-1} \sum_{i=0}^{m} b_{k,j}^j u_i(t) N_j(t) + \sum_{j=0}^{l-1} a_{k,j} N_j(t), \tag{2.80}$$

$$\tilde{f}_k(\tilde{\mathbf{N}}, t, \tau) = \begin{cases} 0 & \text{for } k < l-1, \\ a_{l-1,l} k_1(t-\tau) N_{l-1}(t) & \text{for } k = l-1. \end{cases} \tag{2.81}$$

This general model allows including special cases of multidrug therapy. However, it is valid only in cases when the following conditions are both met:

- each drug affects cells of different type; this is true in the basic model of a killing and a blocking agent treatment, and
- either the molecular source of resistance to each drug is the same (as in multidrug resistance [60, 89]) or the infinite subsystem representing gene amplification is required for only one type of a drug, because resistance to other drugs requires only a single mutation and there is one level of resistance for each of them.

What is more, the finite-dimensional models describing phase-specific chemotherapy can be viewed as special cases of the general model, in which $\tilde{f}_{l-1} = 0$. Moreover, although the antiangiogenic models introduced in the subsequent section cannot be represented in exactly the same way, due to their bilinearity caused by the multiplicative terms containing products of state variables, similar reasoning allows coupling them with drug-resistance models.

Let the system be governed by Eq. (2.53), which afterward is transformed into the form (2.79). The control is bounded:

$$0 \le u_k(t) \le u_{k_{\max}} \le 1, \qquad k = 0, 1, \ldots, m, \tag{2.82}$$

where $u_k(t) = u_{k_{\max}}$ and $u_k(t) = 0$ represent the maximum allowable dose and no application of the k-th drug, respectively. The goal is to minimize the performance index given by (2.23). Due to the particular form of the performance index and the equation governing the model, it is possible to find the solution to the problem by applying a version of Pontryagin maximum principle [97].

Although the performance index (2.23) consists of two components—a sum and an integral—the sum involves another integral, which stems from (2.66)–(2.67). Therefore, it should be rewritten to emphasize this relation:

$$J = \sum_{i=0}^{i=l-1} N_i(T_k) + r_1 N_\Sigma^l(T_k) + \int_0^{T_k} \left[r_1 c_1 N_\Sigma^l(T_k - \tau) N_{l-1}(\tau) + r \sum_{i=0}^m u_i(\tau) d\tau \right].$$

$$\tag{2.83}$$

A number of formulations of necessary conditions for the optimization problem for dynamical systems described by integro-differential equations can be found in the literature [16, 35, 44]. However, they usually are either too general to be efficiently used in such particular problem or have overly strong constraints—for example, smoothness of the control function. Nevertheless, following the reference [16], it is possible to derive the necessary conditions for optimal control:

$$\mathbf{u}^{\mathrm{opt}}(t) = \arg \min_{\mathbf{u}} \left[r \sum_{k=0}^m u_k(t) + \mathbf{p}^T(t)\mathbf{h}(\mathbf{u}, \tilde{\mathbf{N}}) \right.$$
$$\left. + a_{l-1,l} \int_t^{T_k} p_{l-1}(\tau)k_1(t - \tau)N_{l-1}(\tau)d\tau \right], \tag{2.84}$$

$$\dot{\mathbf{p}}^T(t) = - \left[\mathbf{q}^T(t) + \mathbf{p}^T(t)\mathbf{h}_{\tilde{N}}(\mathbf{u}, \tilde{\mathbf{N}}) + \int_t^{T_k} \mathbf{p}^T(\tau)\tilde{\mathbf{f}}_{\tilde{N}}(t - \tau)d\tau \right], \tag{2.85}$$

$$\mathbf{q}(t) = \left[0 \ldots 0 \; r_1 c_1 N_\Sigma^l(T_k - t) \right]^T, \tag{2.86}$$

$$p_i(T_k) = 1, \quad i = 0, 1, \ldots, l-1, \tag{2.87}$$

where $\mathbf{p}(t)$ is a co-state vector.

Since the function $\mathbf{h}(\mathbf{u}, \tilde{\mathbf{N}})$ is bilinear (see (2.80)), the expression in (2.84) is linear with respect to $\mathbf{u}$. This coupled with constraint (2.24) implies that the optimal control has to be of bang-bang type with possible singular arcs.

If singular arcs are not the part of the optimal solution, a gradient method can be applied to find the optimal number of switches and switching times [106, 107].

For example, for the model represented in Fig. 2.11a, and the weighing factors for each cell type set to 1, the necessary conditions are given by [107]:

$$u^{\text{opt}}(t) = \arg\min_{u} \left[p_1(t) \left(\alpha N_{\Sigma}^1 (T - t) N_0(t) + r u(t) \right) \right.$$

$$\left. + 2p_2(t) \left((1 - 2u(t)) \lambda N_0(t) - \alpha N_0(t) \right) + d \, \alpha \int_t^T p_2(\tau) \phi_1(t - \tau) N_0(\tau) d\tau \right],$$

$$(2.88)$$

$$\dot{p}_1(t) = 0, \tag{2.89}$$

$$\dot{p}_2(t) = -\alpha p_1(t) N_{\Sigma}^1 (T - t) + p_2(t) \left[(1 - 2u(t)) \lambda - \alpha \right] + d\alpha \int_t^T p_2(\tau) \phi_1(\tau - t) d\tau,$$

$$(2.90)$$

$$p_1(T) = 1, \tag{2.91}$$

$$p_2(T) = 1, \tag{2.92}$$

Examples of results for that case are shown in Fig. 2.15.

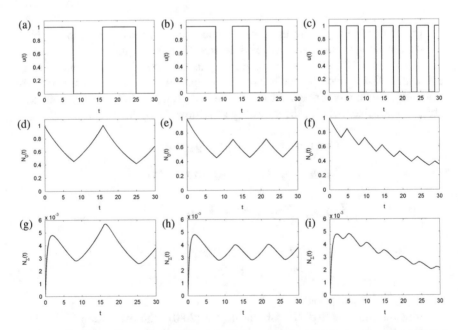

Fig. 2.15 Sample results obtained for the infinite-dimensional model for an assumed number of switches in (**a**),(**b**), (**d**), (**e**), (**g**), (**h**) and periodic control in (**c**), (**f**), (**i**)

The solutions shown in this chapter indicate that switching controls most often provide the best possible results. Moreover, numerical analysis shows that periodic controls are at least suboptimal and, quite often, if there exist better non-periodic solutions, the corresponding value of the performance index is not significantly smaller, compared to periodic solution [77, in Chap. 3]. It is interesting that similar conclusion can be reached when an alternative goal of a therapy is stated as balancing resistant and non-resistant populations [45, 103].

References

1. D.G. Albertson, Gene amplification in cancer. Trends Genet. **22**(8), 447–455 (2006)
2. Z. Agur, The effect of drug schedule on responsiveness to chemotherapy. Ann. N. Y. Acad. Sci. **504**, 274–277 (1988)
3. B. Ainseba, C. Benosman, Optimal control for resistance and suboptimal response in CML. Math. Biosci. **227**(2), 81–93 (2010)
4. R. Airley, *Cancer Chemotherapy* (Wiley-Blackwell, New York, 2009)
5. B. Alberts, A. Johnson, J. Lewis, M. Raff, K. Roberts, P. Walter, *Molecular Biology of the Cell*, 4th edn. (Garland Science, New York, 2002)
6. M.R. Alison, C.E. Sarraf, *Understanding Cancer – From Basic Science to Clinical Practice* (Cambridge University Press, New York, 1997)
7. M. Andreeff, A. Tafuri, P. Bettelheim, P. Valent, E. Estey, R. Lemoli, A. Goodacre, B. Clarkson, F. Mandelli, A. Deisseroth, Cytokinetic resistance in acute leukemia: recombinant human granulocyte colony-stimulating factor, granulocyte macrophage colony- stimulating factor, interleukin-3 and stem cell factor effects in vitro and clinical trials with granulocyte macrophage colony-stimulating factor, in *Haematology and Blood Transfusion 34, Acute Leukemias - Pharmacokinetics*, ed. by Hidemann et al. (Springer, Berlin, 1992), pp. 108–116
8. O. Arino, A survey of structured cell population dynamics. Acta Biotheor. **43**, 3–15 (1995)
9. O. Arino, A. Bertuzzi, A. Gandolfi, E. Sanchez, C. Sinisgali, A model with growth retardation for kinetic heterogeneity of tumor cell populations. Math. Biosci. **206**, 185–199 (2007)
10. D. Axelrod, K. Baggerly, M. Kimmel, Gene amplification by unequal chromatid exchange: probabilistic modeling and analysis of drug resistance data. J. Theor. Biol. **168**, 151–159 (1993)
11. M. Baer, Clinical significance of multidrug resistance in AML: current insights. Clin. Adv. Hematol. Oncol. **3**(12), 910–912 (2005)
12. K. Bahrami, M. Kim, Optimal control of multiplicative control systems arising from cancer therapy. IEEE Trans. Autom. Control **20**, 537–542 (1975)
13. J. Banasiak, M. Lachowicz, M. Moszynski, Chaotic behavior of semigroups related to the process of gene amplification-deamplification with cell proliferation. Math. Biosci. **206**(2), 200–215 (2007)
14. D. Banerje, E. Ercikan-Abali, M. Waltham, B. Schnieders, D. Hochhauser, W. Li, J. Fan, R. Gorlick, E. Goker, J. Bertino, Molecular mechanisms of resistance to antifolates, a review. Acta Biochim. Pol. **42**(4), 457–64 (1995)
15. C. Basdevant, J. Clairambault, F. Levi, Optimisation of time-scheduled regimen for anti-cancer drug infusion. Math. Model. Numer. Anal. **39**(6), 1069–1086 (2005)
16. R. Bate, The optimal control of systems with transport lag. Adv. Control Syst. **7**, 165–224 (1969)
17. R. Bellman, *Mathematical Methods in Medicine* (World Scientific, Singapore, 1983)
18. A. Bertuzzi, A. Gandolfi, R. Vitelli, A regularization procedure for estimating cell kinetic parameters from flow-cytometric data. Math. Biosci. **82**, 63–85 (1986)

19. G. Bonadonna, M. Zambetti, P. Valagussa, Sequential of alternating Doxorubicin and CMF regimens in breast cancer with more then 3 positive nodes. Ten years results. J. Am. Med. Assoc. **273**, 542–547 (1995)
20. D. Borys, R. Jaksik, M. Krzeslak, J. Smieja, A. Swierniak, Cancer - a story on fault propagation in gene-cellular networks, in *Propagation Phenomena in Real World Networks*, ed. by D. Krol, D. Fay, B. Gabrys (Springer, Heidelberg, 2015), pp. 225–256
21. B. Brown, J. Thompson, A rationale for synchrony strategies in chemotherapy, in *Epidemiology*, ed. by D. Lodwig, K.L. Cooke (SIAM, Philadelphia, 1975), pp. 31–48
22. P. Brown, S. Beverly, R. Schimke, Relationship of amplified dihydrofolate reductase genes to double minute chromosomes in unstably resistant mouse fibroblasts cell lines. Mol. Cell. Biol. **1**(12), 1077–1083 (1981)
23. P. Calabresi, P.S. Schein, *Medical Oncology, Basic Principles and Clinical Management of Cancer* (McGraw-Hill, New York, 1993)
24. B. Chabner, D. Longo, *Cancer Chemotherapy and Biotherapy* (Lippincott-Raven, Philadelphia, 1996)
25. B. Chabner, D.L. Longo, *Cancer Chemotherapy and Biotherapy: Principles and Practice* (Lippincott Willians & Wilkins, Philadelphia, 2006)
26. M. Chappell, N. Evans, R. Errington, I. Khan, L. Campbell, R. Ali, K. Godfrey, P. Smith, A coupled drug kinetics-cell cycle model to analyse the response of human cells to intervention by topotecan. Comput. Methods Prog. Biomed. **89**(2), 169–78 (2008)
27. J. Clairambault, P. Michel, B. Perthame, Circadian rhythm and tumour growth. Comptes Rendus Math. Acad. Sci. Paris **342**(1), 17–22 (2006)
28. S. Clare, F. Nahlis, J. Panetta, Molecular biology of breast cancer metastasis. the use of mathematical models to determine relapse and to predict response to chemotherapy in breast cancer. Breast Cancer Res. **2**, 396–399 (2000)
29. D. Clark, *Molecular Biology: Academic Cell Update* (AP Cell Press/Elsevier, Amsterdam, 2010)
30. L. Cojocaru, Z. Agur, A theoretical analysis of interval drug design for cell-cycle-phase-specific drugs. Math. Biosci. **109**, 85–97 (1992)
31. A. Coldman, J. Goldie, A model for the resistance of tumor cells to cancer chemotherapeutic agents. Math. Biosci. **65**, 291–307 (1983)
32. A. Coldman, J. Goldie, A stochastic model for the origin and treatment of tumors containing drug-resistant cells. Bull. Math. Biol. **48**, 279–292 (1986)
33. M. Collins, R. Dedrick, Pharmacokinematics of anticancer drugs, in *Pharmacologic Principles of Cancer Treatment*, ed. by B.A. Chabner (Saunders, Philadelphia, 1982), pp. 77–99
34. L. Coly, D. Van Bekkum, A. Hagenbeek, Enhanced tumor load reduction after chemotherapy induced recruitment and synchronization in a slowly growing rat leukemia model (BNML) for human acute myelonic leukemia. Leuk. Res. **8**, 953–963 (1984)
35. M. Connor, Optimal control of systems represented by differential-integral equations. IEEE Trans. Autom. Control **AC-17**, 164–166 (1972)
36. H. de Jong, Modeling and simulation of genetic regulatory systems: a literature review. J. Comput. Biol. **9**(1), 67–103 (2002)
37. B.F. Dibrov, A. Zhabotinsky, Y.A. Neyfakh, M.P. Orlova, L.I. Churikova, Optimal scheduling for cell synchronization by cycle-phase-specific blockers. Math. Biosci. **66**, 167–185 (1983)
38. B.F. Dibrov, A.M. Zhabotinsky, Y.A. Neyfakh, M.P. Orlova, L.I. Churikova, Mathematical model of cancer chemotherapy. periodic schedules of phase-specific cytotoxic-agent administration increasing the selectivity of therapy. Math. Biosci. **73**, 1–31 (1985)
39. M. Doherty, M. Michael, Tumoral drug metabolism: perspectives and therapeutic implications. Curr. Drug Metab. **4**, 131–149 (2003)
40. Z. Duda, Evaluation of some optimal chemotherapy protocols by using gradient method. Appl. Math. Comput. Sci. **4**, 257–263 (1994)
41. Z. Duda, Numerical solutions to bilinear models arising in cancer chemotherapy. Nonlinear World **4**, 53–72 (1997)

42. H. Eisen, *Mathematical Models in Cell Biology and Cancer Chemotherapy* (Springer, Berlin, 1979)
43. K. Fister, J.C. Panetta, Optimal control applied to cell-cycle-specific cancer chemotherapy. SIAM J. Appl. Math. **60**, 1059–1072 (2000)
44. R. Gabasov, F. Kirilowa, *Qualitative Theory of Optimal Processes* (Nauka, Moscow, 1971)
45. R.A. Gatenby, A.S. Silva, R.J. Gillies, B.R. Frieden, Adaptive therapy. Cancer Res. **69**, 4894–4903 (2009)
46. L. Goellmann, H. Maurer, Theory and applications of optimal control problems with multiple time-delays. J. Ind. Manag. Optim. **10**(2), 413–441 (2014)
47. L. Goellmann, D. Kern, H. Maurer, Optimal control problems with delays in state and control variables subject to mixed control-state constraints. Optim Control Appl. Methods **30**, 341–365 (2009)
48. J. Goldie, A. Coldman, A mathematical model for relating the drug sensitivity of tumors to their spontaneous mutation rate. Cancer Treat. Rep. **63**, 1727–1733 (1979)
49. S. Goutelle, M. Maurin, F. Rougier, X. Barbaut, L. Bourguignon, M. Ducher, P. Maire, The Hill equation: a review of its capabilities in pharmacological modelling. Fundam. Clin. Pharmacol. **22**, 633–648 (2008)
50. W.M. Haddad, V. Chellaboina, Q. Hui, *Nonnegative and Compartmental Dynamical Systems* (Princeton University Press, Princeton, 2010)
51. P. Hahnfeldt, D. Panigraphy, J. Folkman, L. Hlatky, Tumor development under angiogenic signaling: a dynamic theory of tumor growth, treatment response and postvascular dormacy. Cancer Res. **59**, 4770–4778 (1999)
52. L. Harnevo, Z. Agur, The dynamics of gene amplification described as a multitype compartmental model and as a branching process. Math. Biosci. **103**, 115–138 (1991)
53. L. Harnevo, Z. Agur, Drug resistance as a dynamic process in a model for multistep gene amplification under various levels of selection stringency. Cancer Chemother. Pharmacol. **30**, 469–476 (1992)
54. L. Harnevo, Z. Agur, Use of mathematical models for understanding the dynamics of gene amplification. Mutat. Res. **292**(1), 17–24 (1993)
55. O. Hyrien, A. Goldar, Mathematical modelling of eukaryotic DNA replication. Chromosom. Res. **18**(1), 147–161 (2010)
56. O. Hyrien, K. Marheineke, A. Goldar, Paradoxes of eukaryotic DNA replication: MCM proteins and the random completion problem. Bioessays **25**(2), 116–125 (2003)
57. Y. Iwasa, M. Nowak, F. Michor, Evolution of resistance during clonal expansion. Genetics **172**, 2557–2566 (2006)
58. T. Kaczorek, Weakly positive continuous-time linear systems. Bull. Pol. Acad. Sci. **46**, 233–245 (1998)
59. T. Kaczorek, *Positive 1D and 2D Systems* (Springer, London, 2002)
60. J. Kappelmayer, A. Simon, F. Kiss, Z. Hevessy, Progress in defining multidrug resistance in leukemia. Expert Rev. Mol. Diagn. **4**(2), 209–217 (2004)
61. R. Kaufman, P. Brown, R. Schimke, Loss and stabilization of amplified dihydrofolate reductase genes in mouse sarcoma S-180 cell lines. Mol. Cell. Biochem. **1**, 1084–1093 (1981)
62. M. Kim, K. Brahami, K.B. Woo, A discrete-time model for cell-age, size and DNA distributions of proliferating cells, and its application to the movement of the labeled cohort. IEEE Trans. Bio-Med. Eng. **21**, 387–399 (1974)
63. M. Kimmel, D.E. Axelrod, Mathematical models of gene amplification with applications to cellular drug resistance and tumorigenicity. Genetics **125**, 633–644 (1990)
64. M. Kimmel, A. Swierniak, Control theory approach to cancer chemotherapy: benefiting from phase dependence and overcoming drug resistance, in *Tutorials in Mathematical Biosciences III: Cell Cycle, Proliferation, and Cancer*, ed. by A. Friedman. Lecture Notes in Mathematics, Mathematical Biosciences Subseries, vol. 1872 (Springer, Heidelberg, 2006), pp. 185–222
65. M. Kimmel, F. Traganos, Estimation and prediction of cell cycle specific effects of anticancer drugs. Math. Biosci. **80**, 187–208 (1986)

66. M. Kimmel, D. Axelrod, G. Wahl, A branching process model of gene amplification following chromosome breakage. Mutat. Res. **276**(3), 225–239 (1992)
67. M. Kimmel, A. Swierniak, A. Polanski, Infinite-dimensional model of evolution of drug resistance of cancer cells. J. Math. Syst. Estimation Control **8**(1), 1–16 (1998)
68. N. Komarova, D. Wodarz, Effect of cellular quiescence on the success of targeted CML therapy. PLoS One **2**(10), e990 (2007)
69. M. Konopleva, T. Tsao, P. Ruvolo, I. Stiouf, Z. Estrov, C. Leysath, S. Zhao, D. Harris, S. Chang, C. Jackson, M. Munsell, N. Suh, G. Gribble, T. Honda, W. May, M. Sporn, M. Andreef, Novel triterpenoid CDDO-Me is a potent inducer of apoptosis and differentiation in acute myelogenous leukemia. Blood **99**, 326–335 (2002)
70. F. Kozusko, P. Chen, S.G. Grant, B.W. Day, J.C. Panetta, A mathematical model of in vitro cancer cell growth and treatment with the antimitotic agent curacin A. Math. Biosci. **170**(1), 1–16 (2001)
71. A. Krener, The high order maximum principle and its application to singular control. SIAM J. Control Optim. **15**, 256–293 (1977)
72. R. Krishna, L.D. Mayer, Multidrug resistance (MDR) in cancer. Mechanisms, reversal using modulators of MDR and the role of MDR in influencing the pharmacokinetics of anticancer drugs. Eur. J. Pharm. Sci. **11**, 265–283 (2000)
73. J.L. Lebowitz, S.I. Rubinow, A theory for the age and generation time distribution of a microbial population. J. Math. Biol. **1**, 17–36 (1974)
74. U. Ledzewicz, H. Schaettler, Analysis of a cell-cycle specific model for cancer chemotherapy. J. Biol. Syst. **10**(3), 183–204 (2002)
75. U. Ledzewicz, H. Schaettler, Optimal bang-bang controls for a 2-compartment model in cancer chemotherapy. J. Optim. Theory Appl. **114**(3), 609 637 (2002)
76. U. Ledzewicz, H. Schaettler, The influence of PK/PD on the structure of optimal controls in cancer chemotherapy models. Math. Biosci. Eng. **2**, 561–578 (2005)
77. U. Ledzewicz, H. Schaettler, Drug resistance in cancer chemotherapy as an optimal control problem. Discrete Contin. Dyn. Syst. Ser. B **6**(1), 129–150 (2006)
78. U. Ledzewicz, H. Schaettler, Anti-angiogenic therapy in cancer treatment as an optimal control problem. SIAM J. Control Optim. **46**(3), 1052–1079 (2007)
79. U. Ledzewicz, H. Schaettler, Model of maximizing bone marrow with pharmacokinetics. Math. Biosci. **206**(2), 320–342 (2007)
80. F. Levi, U. Schibler, Circadian rhythms: mechanisms and therapeutic implications. Annu. Rev. Pharmacol. Toxicol. **47**, 593–628 (2007)
81. M. Liscovitch, Y. Lavie, Cancer multidrug resistance: a review of recent drug discovery research. Idrugs **5**, 349–355 (2002)
82. D. Luenberger, *Optimization by Vector Space Methods* (Wiley, New York, 1969)
83. R. Martin, Optimal control drug scheduling of cancer chemotherapy. Automatica **28**, 1113–1123 (1992)
84. R.B. Martin, K.L. Teo, *Optimal Control of Drug Administration in Cancer Chemotherapy* (World Scientific, Singapore, 1994)
85. F. Michor, Mathematical models of cancer stem cells. J. Clin. Oncol. **26**(17), 2854–2861 (2008)
86. F. Michor, M. Nowak, Y. Iwasa, Evolution of resistance to cancer therapy. Curr. Pharm. Des. **12**, 261–271 (2006)
87. S. Missailidis (ed.), *Anticancer Therapeutics* (Wiley-Blackwell, Oxford, 2008)
88. R.R. Mohler, *Bilinear Control Processes with Applications to Engineering, Ecology and Medicine* (Academic, New York, 1973)
89. J. Noergaard, L. Olesen, P. Hokland, Changing picture of cellular drug resistance in human leukemia. Crit. Rev. Oncol. Hematol. **50**, 39–49 (2004)
90. L. Norton, R. Simon, Tumor size, sensitivity to therapy, and design of treatment schedules. Cancer Treat. Rep. **61**, 1307–1317 (1977)
91. L. Norton, R. Simon, The Norton-Simon hypothesis revisited. Cancer Treat. Rep. **70**, 41–61 (1986)

92. M. Oklejewicz, E. Destici, F. Tamanini, R. Hut, R. Janssens, G. van der Horst, Phase resetting of the mammalian circadian clock by dna damage. Curr. Biol. **18**, 286–291 (2008)

93. T. Ozben, Mechanisms and strategies to overcome multiple drug resistance in cancer. Fed. Eur. Biochem. Soc. Lett. **580**(12), 2903–2909 (2006)

94. J. Panetta, A mathematical model of drug resistance: heterogeneous tumors. Math. Biosci. **147**, 41–61 (1998)

95. J. Panetta, P. Schaiquevich, V. Santana, C. Stewart, Using pharmacokinetic and pharmacodynamics modeling and simulation to evaluate importance of schedule in topotecan therapy for pediatric neuroblastoma. Clin. Cancer Res. **14**(1), 318–325 (2008)

96. A. Polanski, M. Kimmel, A. Swierniak, Qualitative analysis of the infinite dimensional model of evolution of drug resistance, in *Advances in Mathematical Population Dynamics - Molecules, Cells and Man*, ed. by O. Arino, D. Axelrod, M. Kimmel (World Scientific, Singapore, 1997), pp. 595–612

97. L. Pontryagin, V. Boltyanski, R. Gamkrelidze, E. Mischenko, *Mathematical Theory of Optimal Processes* (Wiley, New York, 1962)

98. C. Roskelley, M. Bissell, The dominance of the microenvironment in breast and ovarian cancer. Cancer Biol. **12**, 97–104 (2002)

99. R. Schimke, Gene amplification in cultured cells. J. Biol. Chem. **263**, 5989–5992 (1988)

100. S.E. Shackney, T.V. Shankey, Cell cycle models for molecular biology and molecular oncology: exploring new dimensions. Cytometry **35**(2), 97–116 (1999)

101. K.G. Shin, R. Pado, Design of optimal cancer chemotherapy using a continuous-time state model of cell kinetics. Math. Biosci. **59**, 225–248 (1982)

102. H.E. Skipper, Historic milestones in cancer biology: a few that are important in cancer treatment (revisited). Semin. Oncol. **6**(4), 506–514 (1979)

103. K. Smallbone, R.A. Gatenby, R.J. Gillies, P.K. Maini, D.J. Gavaghan, Metabolic changes during carcinogenesis: potential impact on invasiveness. J. Theor. Biol. **244**, 703–713 (2007)

104. J. Smieja, Drug resistance in cancer models, in *Handbook of Cancer Models with Application to Cancer Screening, Cancer Treatment and Risk Assessment*, ed. by W.-Y. Tan, L. Hannin (World Scientific, Singapore, 2008), pp. 425–456

105. J. Smieja, A. Swierniak, Different models of chemotherapy taking into account drug resistance stemming from gene amplification. Int. J. Appl. Math. Comput. Sci. **13**, 297–306 (2003)

106. J. Smieja, Z. Duda, A. Swierniak, Optimal control for the model of drug resistance resulting from gene amplification, in *Preprints of 14th World Congress of IFAC*, Beijing (1999), pp. 71–75

107. J. Smieja, A. Swierniak, Z. Duda, Gradient method for finding optimal scheduling in infinite dimensional models of chemotherapy. J. Theor. Med. **3**, 25–36 (2001)

108. G. Stark, Regulation and mechanisms of mammalian gene amplification. Adv. Cancer Res. **61**, 87–113 (1993)

109. G. Swan, Role of optimal control theory in cancer chemotherapy. Math. Biosci. **101**, 237–284 (1990)

110. A. Swierniak, Optimal treatment protocols in leukemia - modeling the proliferation cycle. IMACS Trans. Sci. Comput. **5**, 51–53 (1989)

111. A. Swierniak, Cell cycle as an object of control. J. Biol. Syst. **3**, 41–54 (1995)

112. A. Swierniak, Z. Duda, Some control problems related to optimal chemotherapy - singular solutions. Int. J. Appl. Math. Comput. Sci. **2**, 293–302 (1992)

113. A. Swierniak, Z. Duda, Singularity of optimal control problems arising in cancer chemotherapy. Math. Comput. Model. **19**, 255–262 (1994)

114. A. Swierniak, M. Kimmel, Optimal control application to leukemia chemotherapy protocols design. Zesz. Nauk. Politechniki Slaskiej **74**, 261–277 (1984) (in Polish)

115. A. Swierniak, A. Polanski, Irregularity of optimal control problem in scheduling of cancer chemotherapy. Int. J. Appl. Math. Comput. Sci. **4**, 263–271 (1994)

116. A. Swierniak, A. Polanski, Some properties of TPBVP arising in optimal scheduling of cancer chemotherapy, in *Mathematical Population Dynamics: Analysis of Heterogeneity*, vol. 2, ed. by O. Arino et al. (Wuerz Publishing, Winnipeg, 1995), pp. 359–370

117. A. Swierniak, J. Smieja, Analysis and optimization of drug resistant and phase specific cancer chemotherapy. Math. Biosci. Eng. **2**, 650–670 (2005)

118. A. Swierniak, A. Polanski, M. Kimmel, Optimal control problems arising in cell-cycle-specific cancer chemotherapy. Cell Prolif. **29**(3), 117–139 (1996)

119. A. Swierniak, M. Kimmel, A. Polanski, Infinite dimensional model of evolution of drug resistance of cancer cells. J. Math. Syst. Estimation Control **8**(1), 1–17 (1998)

120. A. Swierniak, A. Polanski, M. Kimmel, A. Bobrowski, J. Smieja, Qualitative analysis of controlled drug resistance model - inverse laplace and semigroup approach. Control Cybern. **28**(1), 61–75 (1999)

121. A. Swierniak, U. Ledzewicz, H. Schaettler, Optimal control for a class of compartmental models in cancer chemotherapy. Int. J. Appl. Math. Comput. Sci. **13**(3), 357–368 (2003)

122. G. Szakacs, J. Paterson, J. Ludwig, C. Booth-Genthe, M. Gottesman, Targeting multidrug resistance in cancer. Nat. Rev. Drug Discov. **5**(3), 219–234 (2006)

123. A. Tafuri, M. Andreeff, Kinetic rationale for cytokine-induced recruitment of myeloblastic leukemia followed by cycle-specific chemotherapy in vitro. Leukemia **4**, 826–834 (1990)

124. Y. Takemura, H. Kobayashi, H. Miyachi, Cellular and molecular mechanisms of resistance to antifolate drugs: new analogues and approaches to overcome the resistance. Int. J. Hematol. **66**(4), 459–477 (1997)

125. R. Tarnawski, K. Skladowski, A. Swierniak, A. Wygoda, A. Mucha, Repopulation of the tumor cells during radiotherapy is doubled during treatment gaps. J. Theor. Med. **2**(4), 297–305 (2000)

126. B. Teicher (ed.), *Cancer Therapeutics* (Humana Press, Totowa, 1997)

127. J. von Foerster, Some remarks on changing populations, in *Kinetics of Cell Proliferation*, ed. by F. Stohlman (Greene & Stratton, New York, 1959), pp. 382–407

128. G.G. Walter, M. Contreras, *Compartmental Modeling with Networks* (Birkhauser, Boston, 1999)

129. G. Webb, *Theory of Nonlinear Age Dependent Population Dynamics* (Marcel Dekker, New York, 1985)

130. G. Webb, Resonance phenomena cell population chemotherapy models. Rocky Mt. J. Math. **20**, 1195–1216 (1990)

131. B. Windle, B. Draper, Y. Yin, S. O'Gorman, G. Wahl, A central role for chromosome breakage in gene amplification, deletion, formation, and amplicon integration. Genes Dev. **5**, 60–174 (1991)

132. H. Wong, L. Vernillet, A. Peterson, J.A. Ware, L. Lee, J.-F. Martini, P. Yu, C. Li, G. Del Rosario, E.F. Choo, K.P. Hoeflich, Y. Shi, B.T. Aftab, R. Aoyama, S.T. Lam, M. Belvin, J. Prescott, Bridging the gap between preclinical and clinical studies using pharmacokinetic-pharmacodynamic modeling: an analysis of GDC-0973, a MEK inhibitor. Clin. Cancer Res. **18**(11), 3090–3099 (2012)

133. L. Zadeh, C. Desoer, *Linear System Theory. The State Space Approach* (McGraw-Hill, New York, 1963)

Chapter 3
Therapy Optimization in Population Dynamics Models

Abstract Cell subpopulations dealt with in the preceding chapter were the result of compartmentalizing the model. In this chapter we focus on growth of cancer cells and the associated vascular system and interactions with the immune system. Therefore, though the variables will still describe the amount of cells of different type, there will be no flux from one type to another. We start with introduction of standard models of population dynamics, with exponential, Gompertzian, and logistic growth. The difference between stochastic and deterministic approach to model population size is explained, a point often misinterpreted in various sources. Then, the angiogenic aspects of cancer growth are discussed, followed by several models of cancer growth including vascularization. Consequently, the problem of optimization of antiangiogenic and combined therapies is analyzed. Finally, models of gene and immunotherapy in cancer treatments are briefly reviewed. In addition to discussion of optimal treatment protocols, the focus of this chapter is on the formal, system engineering-based analysis of dynamical properties of systems under investigation, such as stability and controllability.

3.1 Standard Models of Population Dynamics

The simplest model of population growth is based on the assumption of symmetric cell division that leads to a twofold increase of population size. Therefore, if N_i denotes a number of cells in a given population, the difference equation describing such growth takes the following form:

$$N_{i+1} = 2N_i, \qquad N_0 - \text{given}, \quad i = 1, 2 \dots \tag{3.1}$$

Its solution is given by

$$N_i = 2^i N_0. \tag{3.2}$$

It constitutes an example of geometric (or discrete exponential) growth that is sometimes also called Malthusian growth (after the famous demographer T.R. Malthus (1766–1834), who in 1798 concluded in his *Essay on the Principle of Population*

© Springer International Publishing Switzerland 2016
A. Świerniak et al., *System Engineering Approach to Planning Anticancer Therapies*, DOI 10.1007/978-3-319-28095-0_3

[61] that if no obstacles were encountered by a population, it would increase geometrically, which would inevitably lead to a demographic catastrophe).

Having looked closer at the model (3.1), it becomes clear that it assumes not only no constraints imposed on population but also full synchronization of the cell cycle (all cells exactly at the same phase of the cell cycle) and its constant duration. These assumptions are not satisfied by real cell populations as they exhibit large heterogeneity and are asynchronously distributed over the cell cycle. The simplest way to overcome this simplification is to switch from a discrete to a continuous model in time. Denoting by T_D the doubling time of the population size and introducing a continuous variable $t = iT_D$ representing time, we obtain the following continuous-time model:

$$N(t) = 2^{\frac{t}{T_D}} N_0. \tag{3.3}$$

As a result the following relations hold true:

$$\log_2 \frac{N(t)}{N_0} = \frac{t}{T_D} \tag{3.4}$$

or

$$\ln \frac{N(t)}{N_0} = \frac{\ln 2}{T_D} t. \tag{3.5}$$

Denoting

$$\alpha = \frac{\ln 2}{T_D} \approx \frac{0.639}{T_D} \tag{3.6}$$

leads to the following time-continuous model of Malthusian growth:

$$N(t) = e^{\alpha t} N_0, \tag{3.7}$$

which is a solution to the following linear differential equation:

$$\frac{dN(t)}{dt} \triangleq \dot{N}(t) = \alpha N(t), \qquad N(0) = N_0 \tag{3.8}$$

Such model can be also obtained from a mean-value description of a birth and death process in a cell population:

$$N(t + \Delta t) = N(t) + \gamma N(t)\Delta t - \delta N(t)\Delta t, \tag{3.9}$$

in which it is assumed that the number of both new cells born in the population and those dying in a given period of time is proportional to the number of cells in the population at the beginning of this period. Then, taking the limit $\Delta t \to 0$, we obtain

$$\dot{N} = \lim_{\Delta t \to 0} \frac{N(t + \Delta t) - N(t)}{\Delta t} = (\gamma - \delta)N(t), \tag{3.10}$$

which is the model (3.8) with $\alpha = \gamma - \delta$.

Although choosing continuous time makes it possible to avoid cell cycle synchronization in a whole population, the assumption about constant duration of the cell cycle, expressed in this case in terms of the constant rate of growth, still holds

$$\frac{\dot{N}}{N} = \alpha = \frac{\ln 2}{T_D}. \tag{3.11}$$

In reality, despite the checkpoints that allow transition from one phase of the cell cycle to another being determined in the same way for each cell, the transition times fluctuate. This is particularly evident in the G_1 phase, which contributes the most to the variability of duration of the cell cycle (if one neglects the latent subpopulation G_0). This implies that duration of the cell cycle should be a random variable with mean $\overline{T_D}$. In the simplest case, the random doubling time comes from exponential distribution with a parameter $\tilde{\alpha}$, i.e., the probability distribution function is given by $f(t) = \tilde{\alpha}e^{-\tilde{\alpha}t}$ and the cumulative distribution function by $F(t) = 1 - e^{-\tilde{\alpha}t}$, with $\tilde{\alpha} = \frac{1}{T_D}$. In such case, the population growth model can be applied to describe changes of expected population size in time.

Denoting by dN the change of the mean population size in the infinitesimal time period $(t, t + dt)$, we obtain

$$dN = x^+ dt - x^- dt \tag{3.12}$$

where x^+, x^- represent the mean flows of the newly born and dying cells, respectively. The exponential distribution determines Poisson-type flow of dying cells and, as a result,

$$x^-(t) = \tilde{\alpha}N(t). \tag{3.13}$$

Taking into account that $x^+ = 2x^-$, we obtain

$$\dot{N} = \frac{dN}{dt} = (2\tilde{\alpha} - \tilde{\alpha})N(t) \tag{3.14}$$

and finally we are led to the model identical with (3.8):

$$\dot{N}(t) = \tilde{\alpha}N(t), \tag{3.15}$$

but for the value of the single parameter: $\alpha = \tilde{\alpha}\ln 2$.

Another way to incorporate variable duration of the cell cycle in a model is to assume a deterministic variable determining relative population growth rate α. This leads to the following description of population dynamics:

$$\dot{N}(t) = \alpha(t)N(t) \qquad\qquad N(0) = N_0. \tag{3.16}$$

It is widely known that as the population grows, the growth rate decreases as the cells losing access to nutrients, oxygen, and free space are either more likely to enter apoptosis or slow down cycling. As an example, let us assume that the growth rate α decreases exponentially:

$$\dot{\alpha}(t) = -\beta\alpha(t) \qquad\qquad \alpha(0) = \alpha_0. \tag{3.17}$$

Solving Eqs. (3.16) and (3.17) leads to a saturated growth curve (see Fig. 3.1):

$$N(t) = N_0 e^{\frac{\alpha_0}{\beta}(1-e^{-\beta t})} \tag{3.18}$$

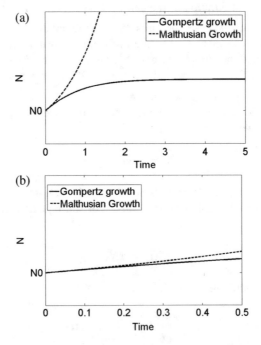

Fig. 3.1 Malthusian and Gompertz growth curves for (**a**) longer and (**b**) shorter time horizon. Initial growth is similar in both models, as illustrated in (**b**)

It can be noticed that

$$\lim_{t \to \infty} N(t) = N_\infty = N_0 e^{\frac{\alpha_0}{\beta}} = N_{\text{max}} \tag{3.19}$$

This curve represents the so-called Gompertz growth, named after the mathematician and actuarial scientist Beniamin Gompertz, who in 1824 postulated an exponential increase of mortality, or death rate, with age [37].

In the particular case that is considered here, the dependence of the rate α on time seems not to be fully justified, as the decrease of population growth rate is related to its size at a given moment and not to its age. However, the Gompertz growth can also be obtained as a solution to the following nonlinear differential equation (called, by the way, a Gompertz equation):

$$\dot{N} = \alpha_0 N - \beta N \ln \frac{N}{N_0} \qquad N(0) = N_0 \tag{3.20}$$

or

$$\dot{N} = -\beta N \ln \frac{N}{N_{\text{max}}}, \tag{3.21}$$

where N_{max} is defined by (3.19). As a result, we are led to the following relation between population size and its growth rate:

$$\frac{\dot{N}}{N} = -\beta \ln \frac{N}{N_{\text{max}}}. \tag{3.22}$$

Though a biological interpretation of the variable growth rate introduced in this particular way is not clear, many experimental growth curves were fitted by the Gompertz growth curve, starting with the pioneering Laird study [53]. Some attempts also have been made to justify Gompertz growth by invoking the existence of the G_0 phase [38].

A logistic curve, which is a solution to another nonlinear differential equation, called Pearl–Verhulst equation, exhibits a very similar shape. The Pearl–Verhulst equation is given by

$$\dot{N} = aN - bN^2 \qquad N(0) = N_0 \tag{3.23}$$

or

$$\dot{N} = aN \left(1 - \frac{N}{N^*_{\text{max}}} \right), \tag{3.24}$$

where $N^*_{\text{max}} = a/b$ and it leads to two equilibrium points: unstable zero point and stable $N = N^*_{\text{max}}$ point. Therefore, N^*_{max} has asymptotically a meaning similar to N_{max} in the Gompertz equation. There is a significant difference between them,

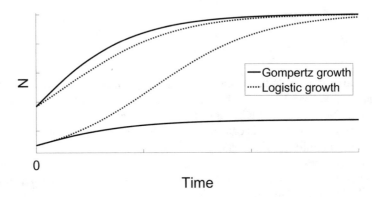

Fig. 3.2 Comparison of Gompertz and logistic growth for a population with the same carrying capacities and initial conditions

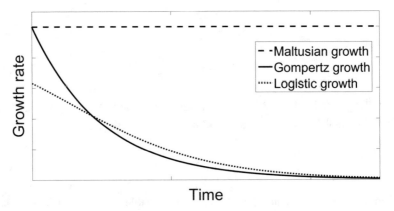

Fig. 3.3 Growth rates as functions of a population size in Malthusian, Gompertz, and logistic models

however: $N_{\max}$ in Gompertzian growth depends on N_0, whereas $N^*_{\max}$ in the logistic growth does not (see also Fig. 3.2). In ecology these values represent the so-called carrying capacities of the environment. It should be noted that not only are growth rates, as functions of time, different in the discussed models, but also the initial (at time 0) growth rate distinguishes logistic from Gompertz growths (Fig. 3.3).

The solution to (3.24) is given by

$$N(t) = \frac{N_0 N^*_{\max} e^{at}}{N^*_{\max} + N_0(e^{at} - 1)}. \tag{3.25}$$

Equation (3.24), introduced in 1838 by P. F. Verhulst [81] as a description of self-limiting growth of a biological population, was rediscovered by R. Pearl in 1925, when he fitted parameters N_0, $N^*_{\max}$ and a to USA and French census data. The logistic curve in either its original or modified form was later found fitting various data on cell populations, in particular bacterial cultures.

There exist other models describing self-limiting growth, in the form of nonlinear differential equations. However, in this book these other descriptions will not be used and therefore they are not discussed. The only exception is a Michaelis–Menten growth model which is introduced in Chap. 5, though in a different context.

3.2 Tumor Angiogenesis and Antiangiogenic Therapy

Angiogenesis is a complex process which leads to formation of new vessels from the existing ones and is stimulated and controlled by molecular factors called activators (stimulators) and inhibitors (blockers). During progression of tumor these factors are released by tumor itself to develop its own vascular network which enables tumor growth and subsequently facilitates cancer metastasis. Since this network is necessary for tumor development, in the seventies of the last century a new anticancer therapy was proposed targeting not directly the cancer cells but the new-born vascular network. This therapy is known as antiangiogenic therapy and the idea is to reduce the tumor volume by reducing its vasculature. It has been first suggested by Folkman [29, 30] more than 40 years ago. The main Folkman's suggestions are as follows:

1. Primary solid tumors proceed through an extended state of avascular growth (almost quiescent) in which maximum attainable size is 1–2 mm in diameter, and the oxygen and nutrients required are supplied by passive diffusion.
2. These microscopic tumors switch on angiogenesis by recruiting surrounding mature host blood vessels to start sprouting new blood vessel capillaries which grow and infiltrate the tumor mass, thus enabling for the growth of the primary tumor and setting the potential for metastatic spread.
3. The angiogenic switch is triggered by production and release of a growth factor (TAF) by tumor cells.
4. Blocking tumor angiogenesis factor or simply destroying newly formed imma-ture blood vessels may be used to affect tumor growth or metastasis.

The most important obstacle against successful chemotherapy is drug resistance acquired by cancer cells while the normal tissues retain sensitivity to the drugs. This negative feature of chemotherapy may be used as an advantage in the antiangiogenic therapy which is directed towards special part of normal tissues and only indirectly destroys tumor cells and it is why it has been called by Kerbel [44] a therapy resistant to drug resistance. Now we know that it is only partially true. Two types of resistance have been observed. The first one, called evasive, includes revascularization as a result of upregulation of alternative pro-angiogenic signals, protection of the tumor and increased metastasis. The second one, called intrinsic, includes rapid adaptive responses, in the case of pre-existing conditions defined by the absence of any beneficial effect of antiangiogenic agents [6]. Nevertheless, therapy directed against tumor vasculature does not exploit tumor cell sensitivity, relying instead on tumor suppression consequent to inhibition of associated vasculature. For more than 10 years Folkman's ideas were not followed by experimental or clinical investigations but now tumor angiogenesis belongs to the most inspiring areas of cancer research in oncology. In [45] ten significant reasons for the explosive growth in tumor angiogenesis research and development of antiangiogenic drugs (we follow the order of points proposed by Kerbel) were presented:

1. Discovery of the basic fibroblast growth factor as the first pro-angiogenic molecule [33].
2. Discovery of the vascular endothelial growth factor and its receptor tyrosine kinases in activated endothelial cells [48].
3. Discovery of angiopoietins and their tyrosine kinase receptors [15].
4. Discovery of endogenous inhibitors of angiogenesis [31].
5. Discovery of additional molecular markers in newly formed blood vessels [8].
6. Development of quantitative assays for angiogenesis [32].
7. Recognition of the prognostic significance of tumor angiogenesis [82].
8. Lack of acquired resistance to directly acting antiangiogenic drugs [45].
9. Discovery of the impact of angiogenesis on liquid hematologic malignancies [14].
10. Discovery of the accidental antiangiogenic effects of various conventional or new anticancer drugs [17].

Complexity of the process of vascularization, as well as the way in which inhibitors, stimulators, and antiangiogenic drugs act, results in complex models (see, e.g., [62]) applicable for simulation of the process but less useful in synthesis or even analysis of therapy protocols. The list of mathematical tools includes partial differential equations (e.g., [42, 63]), stochastic differential equations (e.g., [71, 72]), random walk models [65], cellular automata [2, 3], multi-scale field models [79], and many others (see also Sect. 4.2). The exception is a class of models proposed by Hahnfeldt [39] who suggested that tumor growth with incorporated vascularization mechanism can be described by Gompertz type or logistic type equation with variable carrying capacity which defines the dynamics of the vascular network. Roughly speaking, the main idea of this class of models is to incorporate the spatial aspects of the diffusion of factors that stimulate and inhibit angiogenesis into a non-spatial two-compartmental model for cancer cells and vascular endothelial cells, as opposed to direct modeling of angiogenesis (e.g., [4, 5, 43]). The models considered here belong to this class.

3.3   Models of Cancer Growth Including Vascularization

Incorporating angiogenesis in the idealized scheme of cell cycle kinetics represented by previously considered compartmental models is accomplished by taking into account availability of oxygen and nutrients in cell proliferation. This causes stratification in viability of cells: usually cycling cells are located near the surface or near blood vessels, further layers are occupied by dormant cells, while the deepest regions form a necrotic core. This may lead to self-limiting growth phenomena, which may be described by nonlinear models including logistic or Gompertz-type equations. Before dealing with combined therapy and phase specificity of some drugs, let us first concentrate on how vasculature dynamics can be introduced into the simplest model of growth of cell population.

Let N and K denote volumes occupied by cancer cells and by the vasculature supporting them, respectively, with the dynamics of cell population described by (3.22). Hahnfeldt [39] proposed to treat the carrying capacity which constrains the tumor growth as a varying tumor volume sustainable by the vessels and roughly proportional to the vessel volume, i.e. $N_{max} = K$:

$$\dot{N} = -\beta N \ln \frac{N}{K}.$$

(3.26)

Dynamics of the growth of volume K represented by its potential doubling time, denoted by PDT_k (see (3.17) with $\alpha(t) = 1/PDT_k$) depends on the stimulators of angiogenesis (SF), inhibitory factors secreted by tumor cells (IF) and natural mortality of the endothelial cells (MF):

$$1/PDT_k = MF + SF + IF$$

(3.27)

The spontaneous loss of functional vasculature represented by MF is supposed to be a negative constant, the stimulatory capacity of the tumor upon inducible vasculature represented by SF (angiogenic factors such as vascular endothelial factor) is found to grow at the rate $K^b N^c$ slower than the endogenous inhibition of previously generated vasculature represented by IF (endothelial cell death or disaggregation). It results from the assumption that tumor driven inhibitors from all sites act more systemically whereas tumor-derived stimulators act locally at the individual secreting tumor site. Parameters b and c satisfy the following formula:

$$b + c = 2/3$$

(3.28)

following the assumption that the inhibitor impacts the target endothelial cells in the tumor in a way that ultimately grows as the area of the active surface between the tumor and the vascular network which in turn is proportional to the square of the tumor diameter. It leads to the conclusion that IF is proportional to the tumor volume in power $2/3$ since volume is proportional to the cube of the diameter. The expression for K suggested in [39] has therefore the following form:

$$\frac{\dot{K}}{K} = \frac{\gamma N}{K} - \left(\lambda N^{2/3} + \mu\right),$$

(3.29)

where γ, λ, and μ are constant positive parameters, representing the effect of stimulation, inhibition, and natural mortality, respectively.

The modification of the model given by (3.26) and (3.29), proposed in [19], which also satisfies Hahnfeldt's suggestions, assumes that the effect of SF and MF on the inverse of PDT_K is constant while the IF is proportional to the active surface of the area of tumor being in contact with the vascular network and therefore to the square of the tumor radius:

$$\frac{\dot{N}}{N} = -\beta \ln \frac{N}{K}, \tag{3.30}$$

$$\frac{\dot{K}}{K} = \gamma - \left(\lambda N^{2/3} + \mu\right). \tag{3.31}$$

Combinations of tumor growth models (3.26), (3.30) with vascular network models (3.29), (3.31) result in four nonlinear models of tumor angiogenesis, analyzed in [19, 27, 46, 73].

Application of antiangiogenic therapy can be incorporated in the model using a factor increasing the death rate of the vessels, so that, for example, (3.31) assumes the following form:

$$\frac{\dot{K}}{K} = \gamma - \left(\lambda N^{2/3} + \mu + \eta \cdot u(t)\right). \tag{3.32}$$

where $u(t)$ denotes the dose of the antiangiogenic agent scaled to its effect on vascular network and η is a constant parameter. This factor increases multiplicatively the mortal loss rate of the vessels.

The equilibrium point of the system (3.26)–(3.29), as well as that of (3.30)–(3.31) is given by

$$\tilde{N} = \tilde{K} = \left(\frac{\gamma - \mu}{\lambda}\right)^{3/2} \tag{3.33}$$

or, in case of constant control u^*, if (3.32) is used, by

$$\tilde{N}^* = \tilde{K}^* = \left(\frac{\gamma - \mu - \eta u^*}{\lambda}\right)^{3/2}. \tag{3.34}$$

These equilibrium points exist only when the following conditions are satisfied:

$$\gamma - \mu > 0 \tag{3.35}$$

for (3.33) and

$$\gamma - \mu - \eta u^* \geq 0 \tag{3.36}$$

for (3.34).

These models are strongly nonlinear but it is possible to transform them to a simpler form by making them dimensionless. A logarithmic change of variables

$$x = \ln \frac{N}{N^*}, \qquad y = \ln \frac{K}{\tilde{K}^*}, \qquad \tau = \beta t \qquad (3.37)$$

allows finding asymptotic properties of the equilibria using the standard Lyapunov type analysis of stability (local and global) [74, 78] (see also Appendix A). For example, the modified description of the system (3.32) for a constant control u^* is as follows:

$$\begin{cases} \dot{x} = (y - x), \\ \dot{y} = -\vartheta \left(e^{2/3x} - 1 \right). \end{cases} \qquad (3.38)$$

where $\vartheta = \frac{\gamma - \mu - \eta u^*}{\beta}$. After transformation, the equilibrium point of the system (3.38) is $(x^* = 0, y^* = 0)$ (corresponding to (3.34)). The model linearized around this point is given by

$$\begin{cases} \Delta \dot{x} = (\Delta y - \Delta x), \\ \Delta \dot{y} = -\frac{2}{3}\vartheta \Delta x. \end{cases} \qquad (3.39)$$

From (3.36), it follows that the model (3.39) is asymptotically stable. Therefore, from the first method of Lyapunov (see Appendix A) it follows that the system (3.38) is locally asymptotically stable in the neighborhood of the equilibrium point. In order to check the stability region, the second method of Lyapunov (see Appendix A) can be applied. To do so, it is convenient to introduce yet another variable $z = \beta(y - x)$. Then, (3.38) takes the following form:

$$\begin{cases} \dot{x} = z, \\ \dot{z} = -z - \vartheta \left(e^{2/3x} - 1 \right). \end{cases} \qquad (3.40)$$

The Lyapunov function of Lurie type for this system can be defined as [70, 78]:

$$V(x, z) = \frac{0.5}{\beta} z^2 + (\gamma - \mu - \eta u^*) \int_0^x \left(e^{2/3\xi} - 1 \right) d\xi. \qquad (3.41)$$

Taking into account that $\left(e^{2/3x} - 1 \right) x > 0$ for all $x \neq 0$ and (3.36) is satisfied, the Lyapunov function is positive definite and radially bounded [70]. Its derivative,

$$\dot{V}(x, z) = \frac{\partial V}{\partial x} \cdot \dot{x} + \frac{\partial V}{\partial z} \cdot \dot{z} = -z^2, \qquad (3.42)$$

is negative for all $z \neq 0$. This implies that the equilibrium point is globally asymptotically stable [60, 70] (see also Appendix A). Going back to the original notation of this nontrivial equilibrium point (3.34), it is clear that the constant control u^*, ensuring the asymptotic eradication of the vascular network and, as a result, the tumor, is given by

$$u^* = \frac{\gamma - \mu}{\eta}. \tag{3.43}$$

It was shown in [19] that the same effect might be reached for periodic therapy with mean value satisfying condition (3.43) or greater. However, for periodic control and the original Hahnfeldt model this condition is only necessary and not sufficient since the eradication of the tumor depends on the shape of pulses in the periodic protocol. For other models within this family, this condition is necessary and sufficient. Similar analysis can be carried out for the alternative angiogenesis models [74, 78].

Another class of models of angiogenesis (without control), based on ordinary differential equations, was proposed and analyzed in [1, 4, 34]. The main purpose of these models was to reflect instability of the newly formed vessels structure. Instead of directly coupling tumor and vessel dynamics, as in (3.29) and (3.31), their growth was linked indirectly, through a regulatory protein, whose amount changed as a result of the changing volumes of the vessels and tumor. In addition to introducing an additional, biologically relevant element, the equations describing the system were written in a general way:

$$\begin{cases} \dot{N} = f_1\left(\frac{K}{N}\right) N, \\ \dot{P} = f_2\left(\frac{K}{N}\right) N - \delta P, \\ \dot{K} = f_3(P)K, \end{cases} \tag{3.44}$$

where P denotes the amount of a regulatory protein. Additionally, another model was analyzed in the same work, taking into account the time delay needed before newly produced proteins can increase vasculature volume, as well as time delay that arises between appearance of new vasculature and increasing rate of tumor growth:

$$\begin{cases} \dot{N} = f_1\left(E(t - \tau_1)\right) N, \\ \dot{P} = f_2\left(E\right) N - \delta P, \\ \dot{K} = (f_3(P(t - \tau_2)) - f_1(E(t - \tau_1))) E, \end{cases} \tag{3.45}$$

where $E = \dfrac{K}{N}$.

Such models allow for a relatively easy numerical analysis of various hypotheses concerning specific forms of the functions f_i that might arise from biological considerations. However, it seems that the most important conclusion was drawn from formal stability analysis described in [34]. On one hand, it proved existence of instability and cycles observed in the angiogenesis process. However, the model cannot reproduce the stable behavior observed in less aggressive tumors. This leads to a conclusion that either other processes must be taken into account in analysis or the processes incorporated in the model should be described in another way. Indeed, in [9], a modification of the model implies nontrivial stable equilibrium points. Ability to draw such conclusions is an important advantage of using mathematics in biology, where researchers often do not know, if a key element is not missing in their knowledge about processes that they investigate. In the section on drug resistance it was proved that what was known was sufficient to explain resistance dynamics. Here, in turn, it is shown that by adding new elements to the model (such as time delay between production of regulatory proteins and their effect on vasculature volume), its performance is made more realistic.

3.4 Planning Antiangiogenic and Combined Therapies

Let us consider the model (3.32), in which the control variable u represents therapy. The performance index can be defined again by (2.22), with $N_1 = N$ and $N_2 = K$, and with the isoperimetric constraint (2.21). However, it is more convenient to reformulate both system description and the performance index, using modified variables [78]:

$$x = \ln \frac{N}{\tilde{N}}, \qquad y = \ln \frac{K}{\tilde{K}} \qquad (3.46)$$

It should be noted that the form of (3.46) is different from (3.37) because of using another equilibrium point in the change of variables—(3.33) compared to (3.34), respectively. The system is then described by the following set of equations:

$$\begin{cases} \dot{x} = \beta \cdot (y - x), \\ \dot{y} = -(\gamma - \mu)\left(e^{2/3x} - 1\right) - \eta u. \end{cases} \qquad (3.47)$$

The modified performance index is given by

$$I = r_1 x(T_k) + r_2 y(T_k) + r \int_0^{T_k} u(\tau)d\tau, \qquad (3.48)$$

where r_1, r_2, and r are constant weight factors whose values are chosen depending on the type of therapy used and the strength of the integral constraint. Although

J given by (2.22) and I from (3.48) are not the same indices, and the respective optimization problems are not completely equivalent, they should lead to the same optimal strategies because of the monotonic transformations of variables. The additional term related to the volume of vascular network may be regarded as another constraint imposed on the possible dynamics of the system. On the other hand, by choosing the weight factors we obtain a new possibility of analysis of the mutual dependence between the tumor growth and the volume of the vascular network. Thus, it is reasonable to provide an extensive analysis of their effect on the solution of the optimal control problem. Necessary conditions of optimality can be found using the Pontryagin maximum principle for the following Hamiltonian and co-state variables p_1 and p_2 [78] (see also Appendix B):

$$u_{\mathrm{opt}} = \arg \min_u H = ru + p_1(y - x) + p_2 \left(\eta u - (\gamma - \mu) \left(e^{2/3x} - 1 \right) \right) \qquad (3.49)$$

$$\begin{cases} \dot{p}_1 = p_1 + \frac{2}{3}p_2(\gamma - \mu)e^{2/3x}, \\[2mm] \dot{p}_2 = -p_1, \end{cases} \qquad (3.50)$$

with final conditions

$$p_1(T_f) = r_1, \qquad p_2(T_f) = r_2. \qquad (3.51)$$

Due to linear form of the Hamiltonian with respect to the control variable u, the optimal control is given by

$$u = \begin{cases} 1 \text{ if } r + \eta p_2 < 0, \\[2mm] 0 \text{ if } r + \eta p_2 > 0. \end{cases} \qquad (3.52)$$

The singular control might be the part of the optimal solution for $p_2 = -\dfrac{r}{\eta}$. To check if this is an admissible solution, let us rewrite (3.50) in the following way:

$$\ddot{p}_2 - \dot{p}_2 + \frac{2}{3}(\gamma - \mu)e^{2/3x}p_2 = 0, \qquad (3.53)$$

with $\dot{p}_2(T_f) = -r_1$ and $p_2(T_f) = -r_2$.

Singular arcs are not feasible here, since there are no finite intervals of constant solutions to the co-state equation. This leads to the conclusion that intermediate doses of the drug are not optimal, and that the optimal protocol contains only switches between maximal dose and no drug intervals. Similar conclusion was reached for most modifications of the original Hahnfeldt model [74]. The only model in which singular arcs as parts of optimal trajectory could not be eliminated is the standard Hahnfeldt model with Gompertz-type tumor growth as it has been proved rigorously in [55]. Moreover, in optimization problems whose solutions include singular arcs it appears that suboptimal bang-bang solutions lead to tumor volume dynamics close to the optimal ones while being clinically realizable [57, 58]. An exhaustive review of optimal solutions for a broad class of models from this family of models can be found in [56].

Yet another possibility, arguably more feasible from clinical point of view, is to combine chemotherapy and antiangiogenic therapy [35, 73, 80]. Then, drawbacks of chemotherapy (induced drug resistance, smaller efficiency for slowly growing tumors) could be outweighed by advantages of antiangiogenic therapy and drawbacks of this therapy could be at least slightly moderated by the advantages of chemotherapy. Although tumor eradication in such combined therapy may be still the primary goal, the chaotic structure of the angiogenically created network leads to another target for antiangiogenic agents. Namely using angiogenic inhibitors to normalization of the abnormal vasculature (the so-called pruning effect) facilitate drug delivery [21, 59].

Including chemotherapy action in the model (3.30), (3.55) leads to the following system of equations:

$$\frac{\dot{N}}{N} = -\beta \ln \frac{N}{K} - \psi N v(t),\tag{3.54}$$

$$\frac{\dot{K}}{K} = \gamma - \left(\lambda N^{2/3} + \mu\right) - \eta u(t) - \xi v(t).\tag{3.55}$$

Primary purposes of an individual therapy now differ from these previously stated. It has been already mentioned that the antiangiogenic drug indirectly affects tumor by destabilizing tumor vasculature. On the other hand, cytotoxic agents affect cancer cells directly, but additionally influencing healthy tissue, including endothelial cells. Therefore, the term $v(t)$ has been added to both equations. It denotes the dose of the chemotherapy scaled to its effect on tumor and normal tissues, with ψ and ξ being constant scaling parameters. It is justified even more, since antiangiogenic activity of some chemotoxic agents has been reported [11, 13]. Both $u(t)$ and $v(t)$ might be either constant or periodic functions representing constant drug infusion or periodic administration of the drug, respectively. In some sense both therapies cooperate in achieving common goal, which is tumor and vascular network eradication as $t \to \infty$. From the clinical point of view, it is important to 'push' tumor back to the avascular stage, opening the possibility for surgical intervention.

For constant controls $u(t) = U$ and $v(t) = V$ the equilibrium points of the system (3.54)–(3.55) are

$$N^* = \left(\tfrac{\gamma - \mu - \eta U - \xi V}{\lambda}\right)^{2/3}$$
$$K^* = N^* e^{\frac{\psi V}{\beta}}.\tag{3.56}$$

This leads to the conclusion that the therapy goal as stated above can be achieved for

$$\eta U + \xi V = \gamma - \mu, \tag{3.57}$$

which means that tumor and vascular network eradication might be achieved by application of combined antiangiogenic and chemotherapy with appropriate dosages independently of initial tumor and vasculature sizes.

It should be noted that in this way tumor eradication is achieved, but only asymptotically. Such result is not practical from the clinical point of view, in which a finite therapy horizon should be considered. In such case another concept originating in system theory can be applied—controllability. Since control variables are nonnegative and their maximum values are constrained, we are led to the concept of constrained, or U_C-controllability (for definitions and some fundamental results, see Appendix A).

Usually, controllability conditions are easily derived for linear systems and afterward appropriate mathematical theory is applied to conclude about this property, at least locally, for nonlinear ones. This is the approach that can be used here. After logarithmic transformation of state variables (3.46), scaling of time and parameters, we are led to the following semilinear system:

$$\begin{aligned}
\dot{x} &= y - x - \varepsilon v, \\
\dot{y} &= \vartheta \left(1 - e^{\frac{2}{3}x}\right) - \sigma u - \zeta v,
\end{aligned} \tag{3.58}$$

where $\vartheta = \frac{\gamma - \mu}{\beta}, \sigma = \frac{\eta}{\beta}, \zeta = \frac{\xi}{\beta}$ and time is scaled as $\tau = \beta t$.

The linear system, associated with (3.55)–(3.54), is described by Swierniak and Klamka [75]

$$\dot{z} = \mathbf{A}z + \mathbf{B}[u \quad v]^T, t \in [0, T], \tag{3.59}$$

where z is the state of the associated linear system and

$$\mathbf{A} = \begin{bmatrix} -1 & 1 \\ -\frac{2}{3}\vartheta & 0 \end{bmatrix}, \mathbf{B} = \begin{bmatrix} 0 & -\epsilon \\ \sigma & \zeta \end{bmatrix} \tag{3.60}$$

The admissible control variables are positive, hence the set of admissible controls is a positive cone U_c in the space R^2.

To provide constrained controllability of (3.59), two conditions should be satisfied (Appendix A). First of them involves checking the so-called controllability rank condition:

$$\text{rank}\,[\mathbf{B}|\mathbf{B}\mathbf{A}] = 2. \tag{3.61}$$

The other condition concerns eigenvectors of $\mathbf{A}^T$. There should be no real eigenvector $w \in R^2$ satisfying inequalities $w^T \mathbf{B}[u \quad v]^T \leq 0$ for all $[u \quad v] \in U_c$. The characteristic polynomial of the matrix $\mathbf{A}^T$ is

$$P(s) = s^2 + s + 2/3\vartheta \tag{3.62}$$

Therefore, there are only two cases, when real eigenvectors corresponding to the eigenvalues of $\mathbf{A}^T$ exist. If $\vartheta = 3/8$, the real eigenvector $w = [-1 \quad 2]$ corresponds to the double eigenvalue $s = -0.5$. Then, taking into account that

$$w^T \mathbf{B}[u \quad v]^T = -2\sigma u + (\varepsilon - 2\xi)v > 0$$

for some positive controls, the condition mentioned above is satisfied. If, in turn, $\vartheta < 3/8$, there are two eigenvectors

$$w_1 = \begin{bmatrix} -1 \\ -s_1^{-1} \end{bmatrix}, w_2 = \begin{bmatrix} -1 \\ -s_2^{-1} \end{bmatrix} \tag{3.63}$$

corresponding to the eigenvalues

$$\begin{aligned} s_1 &= 0.5 \left(-1 - \sqrt{1 - 8/3\vartheta} \right), \\ s_2 &= 0.5 \left(-1 + \sqrt{1 - 8/3\vartheta} \right), \end{aligned} \tag{3.64}$$

and it is easy to check that neither of them satisfies the condition $w^T \mathbf{B}[u \quad v]^T \leq 0$ for all $[u \quad v] \in U_c$. Moreover, in these cases the eigenvalues are real and negative. It means that any arbitrary final state can be reached from its neighborhood in finite time using admissible controls, in this case, a combination of antiangiogenic and chemotoxic agents. As mentioned above, a desired final state in the combined therapy may be related not only to tumor eradication but also to normalization of vascular network (see, e.g., [18]).

Constrained controllability (global) of the linear system (3.59) implies local constrained controllability of (3.54) [49, 50]. Hence, the system (3.55)–(3.54) is U_c-locally controllable in $[0, T]$ [50]. If only one of these two forms of therapy is used, however, such outcome may prove to be unattainable. It has been proved in [75] that the property of local constrained controllability is attained only when model parameters satisfy additional conditions related to oscillatory behavior in the untreated case.

Of course, an open question is how to drive the system to the neighborhood of the desired state. One way to realize this objective is to find a solution to a respectively defined optimal control problem which is discussed further in this section.

Annihilation of the vascular network and eradication of the tumor can be also reached by a periodic therapy, which is more reasonable from clinical point of view. The condition for asymptotic eradication of the tumor is then given by Dolbniak and

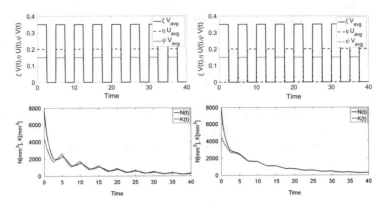

Fig. 3.4 (**a**), (**b**) Two different combined periodic therapies applied to D'Onofrio-Gandolfi model (3.54)–(3.55) and (**c**), (**d**) corresponding state trajectories. Simulation of the drug administration with 5 days intervals on and off therapy

Swierniak [18]

$$\gamma - \mu \le \eta U_{\text{avg}} + \xi V_{\text{avg}}, \tag{3.65}$$

where U_{avg} and V_{avg} denote average drug doses over their periods T_1 and T_2, respectively, defined by

$$U_{\text{avg}} = \frac{1}{T_1} \int_0^{T_1} u(t)dt, \tag{3.66}$$

$$V_{\text{avg}} = \frac{1}{T_2} \int_0^{T_2} v(t)dt \tag{3.67}$$

To illustrate applicability of the results given above, realistic combined periodic therapies were simulated. Figure 3.4 presents sample results. The upper part of Fig. 3.4 presents the dosage of cytotoxic agents and antiangiogenic drug, scaled with parameters η, ξ, ψ. In the bottom part we can observe asymptotic behavior of both populations. Two therapy regimens are compared. In the first one, drugs are being administrated simultaneously for 5 days, followed by drug-free intervals of the same length, the protocol continuing over the 40 days long therapy. As expected, the tumor is shrinking and the carrying capacity of vascular network is decreasing. It is important to notice a significant regrowth of the tumor during drug-free intervals. In the second one, antiangiogenic and chemotherapy were applied in a successive manner, decreasing the bursts of the tumor size.

Optimization of the combined therapy in the finite time horizon seems to be the way of overcoming drawbacks of asymptotic results. As previously, let the performance index combine the final population of cancer cells and the integrals of control variables:

$$\min \leftarrow J = N(T_k) + \int_0^{T_k} (r_u u(t) + r_v v(t))\, dt \tag{3.68}$$

It is implied, once more, from the isoperimetric form of the primary optimization problem in which the integral constraint for chemotherapy represents the feasible cumulated negative effect of the cytotoxic agents, and for the antiangiogenic agent it mostly represents the shortage in the availability of the agent (due to financial constraints) and only in part the possible side effects of the drugs.

Concerning the emergence of drug resistance to tumor chemotherapy, a possible solution might be a model combining one of the previously discussed models of angiogenesis with the simplest model in drug resistance. The simplest model of this type, proposed in [73], consists of three compartments. Population of cancer cells is divided into a sensitive and a resistant subpopulations and the last compartment describes dynamics of supporting vascular network. If the average numbers of cells in the sensitive and the resistant populations are represented by S and R, respectively, the model is given by the following set of equations:

$$
\begin{aligned}
\dot{S} &= -aS + \left(1 - v - \frac{S}{K}\right)(2 - q)aS + rcR \\
\dot{R} &= -cR + (2 - r)\left(1 - \frac{R}{K}\right)cR + (1 - v)qaS \\
\dot{K} &= \gamma N - \lambda N^{2/3}K - \mu K - \eta uK - \xi Kv
\end{aligned}
\tag{3.69}
$$

where $N = S + R$ is a sum of cells from both populations and coefficients a and c stand for the inverses of the average transit times through compartments. Probability of mutations occurring during the process are described by q—probability of mutation into a resistant cell, $0 < q < 1$ and r—probability of mutation into sensitive cell, $0 \le r < 1$. Chemotherapy and antiangiogenic therapy are already incorporated into equations, with v—representing dose of cytotoxic killing agent, $0 \le v \le 1$ and u—representing dose of antiangiogenic drug, $0 \le u \le 1$.

Influence of combined antiangiogenic therapy and chemotherapy on three-compartment systems was tested [18]. Mathematical intricacy provoked generalization of the problem analysis. Instead of deriving necessary and sufficient conditions for tumor eradication, followed by deriving optimality conditions (as it was developed for simpler models, e.g., in [27, 55]), the most promising therapy protocols proposed above for the model (3.54)–(3.55) were checked. Sample simulation results are presented in Fig. 3.5. It appears that for the chosen set of parameters

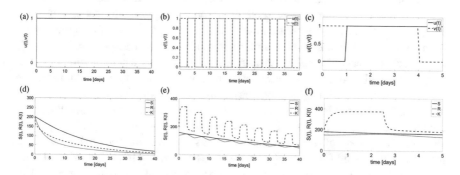

Fig. 3.5 Three-compartment model under (**a**),(**d**) constant, (**b**),(**e**), periodic and (**c**),(**f**) other form of combined therapy for $\eta = 9.1$, $\xi = 4.7$

tumor volume is slightly declining. Population of cells resistant to the action of cytotoxic agents is decreasing, as size of vascular network is decreasing. For such set of scaling parameters we may treat an outcome of the therapy as preventing further expansion of the disease rather than an actual eradication of the tumor. Within the population of endothelial cells we initially observe growth in number of cells, followed by slow declining. A period of 40 days was chosen, during which both cytotoxic and angiogenic agents were administrated to the patient. Outcome of such therapy might be considered satisfying, due to significantly decreased tumor volume and size of vascular network. Nevertheless, it is not recommended to treat a patient with such aggressive therapy regime, since it might be devastating for this patient's overall condition.

Results shown in Fig. 3.5 suggest that in the three-compartmental model a periodic combined therapy yields significantly worse outcome than constant drug infusion. It is in striking contrast to the original Hahnnfeldt model and its modifications, in which the periodic combined therapy theoretically guarantees asymptotic eradication of tumor population [22, 77]. Results of the simulation of application of both agents, with drugs switches occuring every two and a half days, show that as a result also the resistant population is decreasing. At the same time, we observe that the state variable corresponding to the population sensitive to the drug action is slowly declining at a constant rate. As expected, volume of the tumor and size of vascular network decrease; however, the final therapy outcome is poor.

Taking into account requirements related to the size and form of the vascular carrying capacity, it may be convenient to include the final value of K in the performance index, which leads to the functional in the form of

$$J = r_1 N(T_k) + r_2 K(T_k) + \int_0^{T_k} (r_u u(t) + r_v v(t))\, dt$$

or, in the transformed variables,

$$I = r_1 x(T_k) + r_2 y(T_k) + \int_0^{T_k} (r_u u(t) + r_v v(t))\, dt.$$

The first model aimed at optimization of combined anticancer therapy that involved antiangiogenic agents dealt with radiotherapy as the second type of anticancer actions [27]. It was shown that optimal strategies for optimization problem with free final time for a simplified model of angiogenesis combined with LQ model of effects of radiation therapy are constituted by a combination of bang-bang and singular controls. A complete solution in the form of an optimal synthesis for control problem with antiangiogenic treatment only was derived in [54].

Yet another, formally simple modification of the models of antiangiogenic and combined therapy is possible by including time delays in state and/or control variables. The effect of delays in state variables on the asymptotic behavior of the model was discussed first in [20], and, more extensively, in [64]. Two distinct time delays were included in those models: in tumor growth and in vascular capacity. The former represents time needed by a tumor cell population to recognize changes in environment, in terms of reduced nutrient and oxygen availability, due to insufficient vascularity. The latter reflects time lag in the process of vessel growth with respect to stimulus. On the other hand, introducing multiple time delays in control variables enables taking into account differences in PK/PD properties of different drugs (in particular, their half-life times) and clinical recommendation of the order, in which the agents should be administered. Some control properties of such models were discussed in [76].

The analysis presented in this section shows that although the class of models and optimization problems looks very similar to the ones previously analyzed, it leads to the third order system of differential equations, and therefore the synthesis of an optimal control law is much more difficult and remains an open problem.

3.5 Models of Gene and Immunotherapy in Cancer Treatment

Gene therapy and immunotherapy of cancer have been hot topics at least since the beginning of twenty-first century as the therapies that can target cancer cells with none or much less impact on the normal ones. The idea behind gene therapy is to correct what has gone wrong in cancer cells, in one of the following ways:

- providing direct gene correction, especially in the case of mutation of a tumor suppressor gene, or
- suppressing an overexpressed oncogene.

To our best knowledge there are no mathematical models that would describe gene therapy alone. Those that have been published, combines it with immunotherapy. This latter approach, in turn, aims at utilizing the immune system, enhancing its own means to induce a specific anti-tumor response of sufficient size, quality, and duration to produce a clinically meaningful effect. There are three basic ways that can be applied to reach this goal [40]:

- ex vivo transduction of the cytokine genes into tumor cells;
- direct transfer of cytokine genes into tumor cells;
- transfer of tumor antigens or cytokine genes into dendritic cells.

Since viruses of various types are used to achieve the gene transfer, such therapies are often referred to as viral therapies, an interesting review of which can be found in [40].

Additionally, Cytotoxic T Lymphocytes injection can be considered as a way to strengthen natural organizm defenses. Moreover, by means of gene transfer technologies, these can be genetically engineered to become so-called tumor infiltrating T lymphocytes (TILs) expressing a unique high-affinity T cell receptor (TCR) or a chimeric antigen receptor (CAR), both of which confer novel tumor antigen specificity [28]. This is why such approach is often associated with gene therapies. Two distinct approaches can be distinguished: so-called passive and active immunotherapies. Passive immunotherapy is directly aimed at strengthening the potency and precision of the immune system without eliciting a sustained memory response. On the other hand, the goal of active immunotherapy is to provoke the adaptive response that would result in development of immunologic memory. Thus, at least hypothetically, it might be possible to provide sustained protection and reactivity against any recurrence long before a relapse would be clinically apparent [10].

Recently, another approach proved to be very effective, directed at tackling immune suppression, instead of stimulating the immune system. It comes from the realization that cancer takes advantage of mechanisms built into the immune system that act to prevent its overactivation that could harm healthy tissues. These mechanisms are based on receptor-level interactions either during CTL activation (through Cytotoxic T-lymphocyte-associated antigen 4—CTLA4) or CTL actions exerted in peripheral tissues (through programmed death 1 (PD-1) inhibitory receptor) [67]. The concept of delivering antibodies that bind to these receptors, thus inhibiting their inhibitory properties led to the current axioma of "inhibiting the inhibitor" and "breaking tolerance" [25, 26], first proved to be a powerful against melanoma, and then, consequently, against prostate cancer, non-small cell lung carcinoma with agents such as ipilimumab, nivolumab, or lambrolizumab [26, 41, 68].

Furthermore, combined immuno-, chemo- and radiotherapy is under investigation in many research centers. In vitro experiments have demonstrated that pre-treatment of tumor cells with different chemotherapeutic drugs (TAX, DOX, and CIS), can sensitize tumor cells to antigen-specific killing by activated CTLs [36, 66]. Immunotherapy is often associated with viral therapy of cancer, as viruses are used to affect cancer cells and facilitate immune responses. For a recent review of the models, cf. [23].

The simplest model of cancer growth that includes immune system actions consists of two equations describing the dynamics of the tumor cell population N and its interactions with tumor-specific effector cells (i.e., activated CTLs) [52]:

$$\dot{E} = \sigma + \frac{\rho E N}{\eta + N} - \mu E N - \delta E$$
$$\dot{N} = \alpha N(1 - \beta N) - \nu E N \tag{3.70}$$

Here, logistic tumor growth stimulates the production of new effector cells that destroy tumor cells. Three types of steady states are possible in this setting, depending on model parameters ρ, η, μ, δ, α, β, and ν: tumor extinction, tumor growth limited by the carrying capacity of the environment or persistence of a residual tumor controlled by the immune system. Since then, several other models have been proposed, usually tailored to a specific type of cancer that required changes in assumptions about the tumor growth or the form of expressions describing tumor–immune system interactions. For example, in [83] the rate at which effector cells kill tumor cells was limited with respect to the number of cancer cells, whereas (without any justification other than trying to fit clinical data) the effector cells died at a rate proportional to the square of their number. In [69] the system (3.70) is modified into the following form:

$$\dot{E} = \sigma + \frac{\rho EN^c}{\eta + N^c} - \mu EN^c - \delta E$$
$$\dot{N} = \alpha N - \nu EN^c \tag{3.71}$$

with exponential growth replacing the logistic one, an acceptable simplification for tumors at the time of diagnosis. Moreover, to take into account that the contact of tumor and immune system is limited mainly to the tumor surface, the power of c has been introduced into equation. Following these assumptions, it is possible to model combined immuno- and chemotherapy, represented by control variables $i(t)$ and $u(t)$, respectively (immunotherapy here is understood as an injection of CTLs, as in other models described further on)

$$\dot{E} = \sigma + \frac{\rho EN^c}{\eta + N^c} - \mu EN^c - \delta E - k_E Eu(t) + i(t)$$
$$\dot{N} = \alpha N - \nu EN^c - k_N Nu(t) \tag{3.72}$$

where negative effects of chemotherapy on effector cells is also taken into account. In [69] only chemotherapy has been considered and it was assumed to cause loss of both effector and cancer cells, albeit at different rations, represented by the coefficients k_E and k_N, respectively, with

$$u(t) = \begin{cases} 1 \text{ for 1 day after chemotherapy application,} \\ 0 \text{ else.} \end{cases} \tag{3.73}$$

Another similar model of this type is considered in [47]. One should note the resemblance of these equations to those introduced in the section devoted to modeling cancer growth and angiogenesis.

In other models, instead of immune cells that exert direct actions on a cancer population, viruses that induce immune response are directly included. Then, the simplest population model takes the following form [7]:

$$\dot{N_1} = rN_1 \left[1 - (N_1 + N_2)^\epsilon / K^\epsilon \right] - d_v N_1 V - d_N N_1 N_2, \tag{3.74}$$
$$\dot{N_2} = d_v N_1 V - \delta N_2, \tag{3.75}$$
$$\dot{V} = \alpha N_2 - \omega V - d_v N_1 V + u(t), \tag{3.76}$$

where N_1, N_2, and V denote the population size of uninfected and infected cancer cells, and of fee virus, respectively. Parameters r, d_v, d_N, α, and ω are rate constants, K is the carrying capacity, and the parameter ϵ is introduced to account more precisely for the shape of the sigmoidal growth curve. In an apparent violation of the flux balances, the term $d_N N_1 N_2$, describing the rate at which uninfected cells become infected through fusion with infected cells, does not appear in (3.75) because no new members are added to population $N_2(t)$ when an uninfected cell is incorporated into a syncytium. Variable $u(t)$, not appearing explicitly in the original paper, has been put here as a control variable, representing the rate of virus injection. Usually, such control is calculated under assumption that injections correspond to impulse functions, which means that the search for an optimal solution is reduced to finding optimal time instances for injections and the amounts of virus injected.

More detailed models have also been developed and analyzed. In addition to the CD8T cells, they take into account also other cells of the immune system as well as chemotherapeutical agents and PK effects with respect to control variables considered (see, e.g., [12, 16]). Including different types of immune cells in the model facilitates taking into account immune memory, leading to active immunotherapy modeling. For example, in [24], the model includes effector (and effector-memory) cells in the lymphoid compartment Y_{el} and in the periphery Y_{ep}, central memory cells Y_{cm} in the lymphoid compartment, viral loads for the oncolytic virus V_v and the vaccine virus V_a as well as infected and uninfected tumor cells, denoted by N_1 and N_2, respectively:

$$\dot{N}_1 = rN_1 \left[1 - (N_1 + N_2)\right] - d_v \frac{N_1 V_v}{h_u + N_1} - d_u \frac{N_1 Y_{ep}}{h_{ep} + Y_{ep}}, \tag{3.77}$$

$$\dot{N}_2 = d_v \frac{N_1 V_v}{h_u + N_1} - \delta N_2 - d_i \frac{N_2 Y_{ep}}{h_{ep} + Y_{ep}}, \tag{3.78}$$

$$\begin{aligned} \dot{Y}_{cm} &= i_{nc} + Y_{cm} \left(i_c + p_c^a \frac{V_a}{h_a + V_a} + p_c^v \frac{V_v}{h_v + V_v}\right)(1 - k_c Y_{cm}) \\ &+ m_{pl} \frac{Y_{ep}}{1 + g_v V_v} - r_{cl} \frac{V_v Y_{cm}}{h_v + V_v} - d_c Y_{cm} - l_c L(t) Y_{cm} V_v, \end{aligned} \tag{3.79}$$

$$\begin{aligned} \dot{Y}_{el} &= i_{nl} + Y_{el} \left(p_e^a \frac{V_a}{h_a + V_a} + p_e^v \frac{V_v}{h_v + V_v}\right)(1 - k_e Y_{el}) \\ &+ r_{cl} \frac{V_v Y_{cm}}{h_v + V_v} - d_1 Y_{el} - m_{1p} Y_{el} - l_l L(t) Y_{el} V_v, \end{aligned} \tag{3.80}$$

$$\dot{Y}_{ep} = m_{1p} Y_{el} - d_p Y_{ep} - d_t N_1 Y_{ep} - m_{pl} \frac{Y_{ep}}{1 + g_v V_v} - l_1(t) Y_{ep} V_v, \tag{3.81}$$

$$\dot{V}_v = \alpha b N_2 - \omega_v V_v + u_v(t), \tag{3.82}$$

$$\dot{V}_a = -\omega_a V_a + c_a u_a(t). \tag{3.83}$$

There are two control variables, $u_a(t) = u_a \delta(t)$, $u_v(t) = u_v \delta(t - t_0)$, representing injection of oncolytic virus and a vaccine virus, respectively, ($\delta(t)$ is the Kronecker delta function, u_a and u_v are constants), and $L(t) = e^{-q\|t - t_0\|}$.

The analysis of the model (3.77)–(3.83) led to the important conclusion that multi-stability arising in it, associated with transitions between a tumor-free and a tumor-present state, is driven by the immune response, while the multi-instability is driven by the presence of the virus. Moreover, a possible feedback loop caused by the oncolytic virus has been hypothesized as a possible factor limiting the efficiency of the immunotherapy, as it can stop the intra-tumor blood flow and kill the uninfected tumor cells, which, in turn, may limit the spread and persistence of the virus. Therefore, using another type of an oncolytic virus (with a better half-life or a better replication rate) has been suggested to improve therapy outcome.

Similar models were used to solve optimization problems in scheduling immunotherapy protocols. They vary in the number and type of components included in the model (various types of immune cells, cytokines, cancer subpopulations); nevertheless, they provide a valuable insight into how, if at all, the efficiency of existing protocols can be improved.

In [51] treatment of glioblastoma with alloreactive CTL injections was investigated. The model comprised dynamics of tumor cells, CTLs, two types of signalling molecules (IFN-γ and TGF-β cytokines), and two types of major histocompatibility complexes (MHC class I and II). The analysis suggests that tumor eradication requires a 20-fold higher dose than had been administered in clinical studies.

In [47] one of the simplest possible models of combined immuno- and chemotherapy was analyzed. It describes the growth of tumor and immune cells as well as chemotherapeutic drug concentration. Contrary to more detailed models, the immune cells in the model are treated as one group called the effector cells. The search for the best therapy protocol there is stated as a Pareto-optimization problem, with three objective functionals, describing minimal tumor size at the end of therapy, maximum possible tumor size during treatment, and cumulative tumor size over the treatment period. The results obtained by means of a genetic algorithm indicated that the standard protocols might be significantly improved if the immunogenicity factor was taken into account prior to every therapeutic intervention. A threshold number of immunotherapeutic interventions was found that allows to eradicate tumor cells without inflicting unnecessary therapeutic burden with negligible improvement in patient condition.

In turn, in [12], in addition to cancer cells exposing the tumor-associated-antigen (TAA) and the signaling IL-2 cytokine, three classes of immune cells are considered—CD4 T helper cells, tumor-specific CD8 cytotoxic T cells, and mature dendritic cells loaded with the TAA, whose injection constitutes control action. The cost functional to be minimized has been defined, similarly as in Sect. 2.2.4, as the sum of the tumor mass at the end of the therapy and the total quantity of vaccine injected. The integral of the tumor mass exceeding a certain threshold was introduced as an additional component of the functional and the admissible set of control functions was restricted to impulse functions. It was found that the first vaccination using a high dose should be applied at the beginning of the treatment period, while the other vaccinations containing smaller doses should be distributed approximately evenly over the rest of the treatment period.

In [16] a similar model was considered, with additional chemotherapeutic drug action. Although it does not involve therapy optimization, a bifurcation analysis performed there showed that the long-term behavior of the system can be very sensitive to the initial conditions. Since both the zero- and high tumor equilibrium points have been proved to be stable, even very small changes in either the initial tumor size may have drastic consequences for the outcome of the disease. This, in turn, implies that a combination of immuno- and chemotherapy is required to guarantee a stable disease-free state.

References

1. Z. Agur, L. Arakelyan, P. Daugulis, Y. Ginosar, Hopf point analysis for angiogenesis models. Discrete Continuous Dyn. Syst. Ser. B **4**, 29–38 (2004)
2. T. Alarcon, H. Byrne, P. Maini, J. Panovska, Mathematical modelling of angiogenesis and vascular adaptation. Stud. Multidiscip. **3**, 369–387 (2006)
3. A.R.A. Anderson, M.A.J. Chaplain, Continuous and discrete mathematical models of tumor-induced angiogenesis. Bull. Math. Biol. **60**, 857–900 (2003)
4. L. Arakelyan, V. Vainstein, Z. Agur, A computer algorithm describing the process of vessel formation and maturation, and its use for predicting the effects of antiangiogenic and antimaturation therapy on vascular tumor growth. Angiogenesis **5**, 203–214 (2002)
5. A.L. Bauer, T.L. Jackson, Y. Jiang, A cell-based model exhibiting branching and anastomosis during tumor-induced angiogenesis. Biophys. J. **92**(9), 3105–3121 (2007)
6. G. Bergers, D. Hanahan, Modes of resistance to antiangiogenic therapy. Nat. Rev. Cancer **8**, 592–603 (2008)
7. M. Biesecker, J.-H. Kimn, H. Lu, D. Dingli, Z. Bajzer, Optimization of virotherapy for cancer. Bull. Math. Biol. **72**, 469–489 (2010)
8. J. Bischoff, Approaches to studying cell adhesion and angiogenesis. Trends Cell Biol. **5**, 69–73 (1995)
9. M. Bodnar, U. Forys, Angiogenesis model with carrying capacity depending on vessel density. J. Biol. Syst. **17**(1), 1–25 (2009)
10. J.B. Brayer, J. Pinilla-Ibarz, Developing strategies in the immunotherapy of leukemias. Cancer Control **20**(1), 49–59 (2013)
11. T. Browder, C.E. Butterfield, B.M. Kraling, B. Shi, B. Marshall, M.S. OReilly, J. Folkman, Antiangiogenic scheduling of chemotherapy improves efficacy against experimental drug-resistant cancer. Cancer Res. **60**, 1878 (2000)
12. F. Castiglione, B. Piccoli, Cancer immunotherapy, mathematical modeling and optimal control. J. Theor. Biol. **247**, 723–732 (2007)
13. M. Colleoni, A. Rocca, M.T. Sandri, L. Zorzino, G. Masci, F. Nole, G. Peruzzotti, C. Robertson, L. Orlando, S. Cinieri, F. de Braud, G. Viale, A. Goldhirsch, Low-dose oral methotrexate and cyclophosphamide in metastatic breast cancer: antitumour activity and correlation with vascular endothelial growth factor levels. Ann. Oncol. **13**, 73 (2002)
14. R.J. D'Amato, M.S. Loughnan, E. Flynn, J. Folkman, Thalidomide is an inhibitor of angiogenesis. Proc. Natl. Acad. Sci. U.S.A. **91**, 4082–4085 (1994)
15. S. Davis, G.D. Yancopoulos, The angio-poietins: Yin and Yang in angiogenesis. Curr. Top. Microbiol. Immunol. **237**, 173–185 (1999)
16. L.G. de Pillis, W. Gua, A.E. Radunskaya, Mixed immunotherapy and chemotherapy of tumors: modeling, applications and biological interpretations. J. Theor. Biol **238**, 841–862 (2006)
17. J. Denekamp, Angiogenesis, neovascular proliferation and vascular pathophysiology as targets for cancer therapy. Br. J. Radiol. **66**, 181–196 (1993)
18. M. Dolbniak, A. Swierniak, Comparison of simple models of periodic protocols for combined anticancer therapy. Comput. Math. Methods Med. **2013**, 567213 (2013)
19. A. d'Onofrio, A. Gandolfi, Tumor eradication by antiangiogenic therapy: analysis and extensions of the model by Hahnfeldt et al. (1999). Math. Biosci. **191**(2), 159–184 (2004)
20. A. d'Onofrio, A. Gandolfi, A family of models of angiogenesis and anti-angiogenesis anti-cancer therapy. Math. Med. Biol. **26**, 63–95 (2009)
21. A. d'Onofrio, A. Gandolfi, Chemotherapy of vascularised tumours: role of vessel density and the effect of vascular "pruning". J. Theor. Biol. **264**, 253–265 (2010)
22. A. d'Onofrio, U. Ledzewicz, H. Maurer, H. Schaettler, On optimal delivery of combination therapy for tumors. Math. Biosci. **222**, 13–26 (2009)
23. R. Eftimie, J.L. Bramson, D.J.D. Earn, Interactions between the immune system and cancer: a brief review of non-spatial mathematical models. Bull. Math. Biol. **73**, 2–32 (2011)

24. R. Eftimie, J. Dushoff, B.W. Bridle, J.L. Bramson, D.J.D. Earn, Multi-stability and multi-instability phenomena in a mathematical model of tumor-immune-virus interactions. Bull. Math. Biol. **73**, 2932–2961 (2011)

25. A. Eggermont, G. Kroemer, L. Zitvogel, Immunotherapy and the concept of a clinical cure. Eur. J. Cancer **49**, 2965–2967 (2013)

26. A. Eggermont, C. Robert, J.C. Soria, L. Zitvogel, Harnessing the immune system to provide long-term survival in patients with melanoma and other solid tumors. OncoImmunology **3**, e27560 (2014)

27. A. Ergun, K. Camphausen, L. Wein, Optimal scheduling of radiotherapy and angiogenic inhibitors. Bull. Math. Biol. **65**, 407–424 (2003)

28. M. Essand, A.S.I. Loskog, Genetically engineered T cells for the treatment of cancer. J. Intern. Med. **273**, 166–181 (2013)

29. J. Folkman, Tumor angiogenesis: therapeutic implications. N. Engl. J. Med. **295**, 1182–1186 (1971)

30. J. Folkman, Antiangiogenesis: new concept for therapy of solid tumors. Ann. Surg. **175**, 409–416 (1972)

31. J. Folkman, Angiogenesis inhibitors generated by tumors. Mol. Med. **1**, 120–122 (1995)

32. J. Folkman, C. Haudenschild, Angiogenesis in vitro. Nature **288**, 551–555 (1980)

33. J. Folkman, M. Klagsburn, Angiogenic factors. Science **235**, 442–447 (1987)

34. U. Forys, Y. Kheifetz, Y. Kogan, Critical point analysis for three variable cancer angiogenesis models. Math. Biosci. Eng. **2**, 511–525 (2005)

35. G. Gasparini, R. Longo, M. Fanelli, B.A. Teicher, Combination of antiangiogenic therapy with other anticancer therapies:results, challenges, and open questions. J. Clin. Oncol. **23**, 1295–1311 (2005)

36. S. Gill, M. Kalos, T cell-based gene therapy of cancer. Transl. Res. **161**(4), 365–379 (2013)

37. B. Gompertz, On nature of the function expressive of the law of human mortality, and a new mode of determining the value of life contingencies, Letter to F. Batly. Esq. Phil. Trans. R. Soc. **115**, 513–585 (1825)

38. M. Gyllenberg, G.F. Webb, Quiescence as an explanation of Gompertzian tumor growth. Growth Dev. Aging **53**, 25–33 (1989)

39. P. Hahnfeldt, D. Panigraphy, J. Folkman, L. Hlatky, Tumor development under angiogenic signaling: a dynamic theory of tumor growth, treatment response and postvascular dormacy. Cancer Res. **59**, 4770–4778 (1999)

40. K.J. Harrington, R.G. Vile, H.S. Pandha (eds.), *Viral Therapy of Cancer* (Wiley, Hoboken, 2008)

41. F.S. Hodi et al., Improved survival with ipilimumab in patients with metastatic melanoma. N. Engl. J. Med. **363**(8), 711–723 (2010)

42. T. Jackson, X. Zheng, A cell-based model of endothelial cell migration, proliferation and maturation during corneal angiogenesis. Bull. Math. Biol. **72**, 830–868 (2010)

43. R.K. Jain, R.T. Tong, L.L. Munn, Effect of vascular normalization by antiangiogenic therapy on interstitial hypertension, peritumor edema, and lymphatic metastasis: insights from a mathematical model. Cancer Res. **67**(6), 2729–2735 (2007)

44. R. Kerbel, A cancer therapy resistant to resistance. Nature **390**, 335–340 (1997)

45. R. Kerbel, Tumor angiogenesis: past, present and near future. Carcinogenesis **21**, 505–515 (2000)

46. M. Kimmel, A. Swierniak, Control theory approach to cancer chemotherapy: benefiting from phase dependence and overcoming drug resistance, in *Tutorials in Mathematical Biosciences III: Cell Cycle, Proliferation, and Cancer*, ed. by A. Friedman. Lecture Notes in Mathematics. Mathematical Biosciences Subseries, vol. 1872 (Springer, Heidelberg, 2006), pp. 185–222

47. K.L. Kiran, S. Lakshminarayanan, Optimization of chemotherapy and immunotherapy: in silico analysis using pharmacokinetic-pharmacodynamic and tumor growth models. J. Process Control **23**, 396–403 (2013)

48. M. Klagsburn, S. Soker, VEGF/VPF: the angiogenesis factor found? Curr. Biol. **3**, 699–702 (1993)

49. J. Klamka, *Controllability of Dynamical Systems* (Kluwer Academic, Dordrecht, 1991)
50. J. Klamka, A. Swierniak, Controllability of a model of combined anticancer therapy. Control Cybern. **42**(1), 123–138 (2013)
51. N. Kronik, Y. Kogan, V. Vainstein, Z. Agur, Improving alloreactive CTL immunotherapy for malignant gliomas using a simulation model of their interactive dynamics. Cancer Immunol. Immunother. **57**, 425–439 (2008)
52. V.A. Kuznetsov, I.A. Makalkin, M.A. Taylor, A.S. Perelson, Nonlinear dynamics of immunogenic tumors: parameter estimation and global bifurcation analysis. Bull. Math. Biol. **56**(2), 295–321 (1994)
53. A.K. Laird, Cell fractionation of normal and malignant tissues. Exp. Cell Res. **6**(1), 30–44 (1954)
54. U. Ledzewicz, H. Schaettler, Antiangiogenic therapy in cancer treatment as an optimal control problem. SIAM J. Control Optim. **46**, 1052–1079 (2007)
55. U. Ledzewicz, H. Schaettler, Optimal and suboptimal protocols for a class of mathematical models of tumor anti-angiogenesis. J. Theor. Biol **252**, 295–312 (2008)
56. U. Ledzewicz, H. Schaettler, On the optimality of singular controls for a class of mathematical models for tumor antiangiogenesis. Discrete Contin. Dyn. Syst. Ser. B **11**, 691–715 (2009)
57. U. Ledzewicz, J. Marriott, H. Maurer, H. Schaettler, Realizable protocols for optimal administration of drugs in mathematical models for anti-angiogenic treatment. Math. Med. Biol. **27**, 157–179 (2010)
58. U. Ledzewicz, H. Maurer, H. Schaettler, Minimizing tumor volume for a mathematical model of anti-angiogenesis with linear pharmacokinetics, in *Recent Advances in Optimization and Its Applications in Engineering* (Springer, New York, 2010), pp. 267–276
59. J. Ma, D.J.Waxman, Combination of antiangiogenesis with chemotherapy for more effective cancer treatment. Mol. Cancer Ther. **7**, 3670–3684 (2008)
60. M. Malisoff, F. Mazenc, *Constructions of Strict Lyapunov Functions*. Communications and Control Engineering (Springer, London, 2009)
61. T.R. Malthus, *An Essay on the Principle of Population* (St. Paul's Church-yard, London, 1798)
62. N. Mantzaris, S. Webb, H. Othmer, Mathematical modeling of tumor-induced angiogenesis. J. Math. Biol. **49**, 111–127 (2004)
63. S.R. McDougall, A.R.A. Anderson, M.A.J. Chaplain, J.A. Sheratt, Mathematical modelling of flow through vascular networks: implications for tumor-induced angiogenesis and chemotherapy strategies. Bull. Math. Biol. **64**, 673–702 (2002)
64. M. J. Piotrowska, U. Forys, Analysis of the Hopf bifurcation for the family of angiogenesis models. J. Math. Anal. Appl. **382**(1), 180–203 (2011)
65. M.J. Plank, B.D. Sleeman, A reinforced random walk model of tumour angiogenesis and antiangiogenic strategies. Math. Med. Biol. **20**, 135–181 (2003)
66. R. Ramakrishnan, D.I. Gabrilovich, Novel mechanism of synergistic effects of conventional chemotherapy and immune therapy of cancer. Cancer Immunol. Immunother. **62**, 405–410 (2013)
67. A. Ribas, Tumor immunotherapy directed at PD-1. N. Engl. J. Med. **366**(26), 2517–2519 (2012)
68. C.R. Robert et al., Ipilimumab plus dacarbazine for previously untreated metastatic melanoma. N. Engl. J. Med. **364**(26), 2517–2526 (2011)
69. K. Roesch, D. Hasenclever, M. Scholz, Modelling lymphoma therapy and outcome. Bull. Math. Biol. **76**, 401–430 (2014)
70. J.L. Salle, S. Lefschetz, *Stability by Liapunov's Direct Method* (Academic, New York, 1961)
71. B.D. Sleeman, M. Hubbard, P.F. Jones, The foundations of an unified approach to mathematical modelling of angiogenesis. Int. J. Adv. Eng. Sci. Appl. Math. **1**, 43–52 (2009)
72. C.L. Stokes, D.A. Lauffenberger, Analysis of the roles of microvessel endothelial cell random motility and chemotaxis in angiogenesis. J. Theor. Biol. **152**, 377–403 (1991)
73. A. Swierniak, Direct and indirect control of cancer populations. Bull. Pol. Acad. Sci. **56**(4), 367–378 (2008)

74. A. Swierniak, Comparison of six models of antiangiogenic therapy. Appl. Math. **36**(2), 333–348 (2009)
75. A. Swierniak, J. Klamka, Control properties of models of antiangiogenic therapy, in *Advances in Automatics and Robotics (Postepy Automatyki i Robotyki)*, ed. by K. Malinowski, R. Dindorf. Monograph of Committee of Automatics and Robotics PAS, Kielce, 2011, 16, part 2, pp. 300–312
76. A. Swierniak, J. Klamka, Local controllability of models of combined anticancer therapy with delays in control. Math. Model. Nat. Phenom. **9**(4), 216–226 (2014)
77. A. Swierniak, K. Ploskonski, Periodic control of antiangiogenic and combined anticancer therapies, in *Proceedings of the IFAC Workshop on Periodic Control Systems PSYCO 2010*, Antalya, 2010. CD ROM edition
78. A. Swierniak, J. Smieja, Singularity of optimal antiangiogenic strategies - exception or rule, in *Proceedings of the 7th IASTED Biomedical Engineering*, Innsbruck, 2010
79. R.D.M. Travaso, E. Corvera Poire, M. Castro, J.C. Rodriguez-Manzaneque, A. Hernandez-Machado, Tumor angiogenesis and vascular patterning: a mathematical model. Plos ONE **6**, e19989 (2011)
80. L.S. Tseng, K.T. Jin, K.F. He, H.H. Wang, J. Cao, D.C. Yu, Advances in combination of antiangiogenic agents targeting VEGF-binding and conventional chemotherapy and radiation for cancer treatment. J. Chin. Med. Assoc. **73**, 281–288 (2010)
81. P.F. Verhulst, Notice sur la loi que la population poursuit dans son accroissement. Corresp. Math. Phys. **10**, 113–121 (1838)
82. N. Weidner, Intramural microvessel density as a prognostic factor in cancer. Am. J. Pathol. **147**, 9–19 (1995)
83. A.L. Woelke, M.S. Murgueitio, R. Preissner, Theoretical modeling techniques and their impact on tumor immunology. Clin. Dev. Immunol. (2010). doi:10.1155/2010/271794

Chapter 4
Structured Models and Their Use in Modeling Anticancer Therapies

Abstract Tumorigenesis is a very complex pathological process, evolving through different parallel pathways. The list of hallmark capabilities which cancer has to acquire was presented in two famous review papers by Hanahan and Weinberg (Cell 100:57–70, 2000; Cell 144:646–674, 2011). Following recent biological discoveries, especially those in molecular biology, mathematicians try to create models adequate to knowledge in the biomedical field, oriented on specific aspects of tumor development. They apply various modeling techniques in order to perform this task. Among these techniques one can distinguish models based on partial differential equations, single-cell-based models, cellular automata, and others. This chapter is devoted to models with structure. Structure may have different meanings, it may refer to the space where cells develop, cellular level of differentiation, or some other physiological feature of the cell, it may also refer to mutual relations among the elements forming the whole system, as it is in agent-based models.

4.1 Tumor Growth and Treatment in Two and Three Dimensions

The area of mathematical modeling of tumor growth has been widely discussed in the literature. Publications may be classified according to the phase of tumor development which the model describes; they concern hyperplastic and benign tumors [12, 93], tumors in the invasive stage [35, 69], tumors with already induced angiogenesis [34], or tumors in the metastatic stage [138]. Models can also be classified by the mathematical description or the methodology employed. Different approaches include deterministic or stochastic description of the model; some hybrid models may include various approaches; some may be built on the basis of cellular automata [40] (see Sect. 4.6), some on the basis of structured models [43] (see Sect. 4.5); and others on the basis of kinetic theory [16].

Tumor formation is a multistage process [60, 61], and scientists often concentrate only on particular processes in order to analyze them accurately. Thus one can find works devoted solely to, for example, avascular growth of the tumor, growth of tumor interacting with stroma, or metabolism of the tumor in which the invasion is acid-mediated [52].

© Springer International Publishing Switzerland 2016
A. Świerniak et al., *System Engineering Approach to Planning Anticancer Therapies*, DOI 10.1007/978-3-319-28095-0_4

Mathematical models describing the dynamics of the cell population are often based on the ordinary differential equations. Compartmental models are an example. However these latter do not take into account important spatial dependencies. Due to the fact that most of the processes in the living tissues have spatial nature, a large number of mathematical models have been developed. They take into account not only temporal but also spatial changes of the cellular population.

One of the most popular is the continuum approach. It originates from the general conservation equation for the tumor cell density n

$$\frac{\partial n}{\partial t} + \nabla \cdot \mathbf{J} = f(n), \tag{4.1}$$

where ∇ is a vector differential operator (nabla operator), $\mathbf{J}$ denotes flux of the cells, and $f(\cdot)$ is a reaction term. By inserting advective and diffusive fluxes to the continuity equation, we arrive at the equation with a transport term on the left-hand side and diffusion term on the right-hand side

$$\frac{\partial n}{\partial t} + \nabla \cdot (\mathbf{v}n) = \nabla \cdot (D\nabla n) + f(n), \tag{4.2}$$

where $\mathbf{v}$ stands for velocity of the transport movement, and D stands for the diffusion coefficient.

4.1.1 Chemotaxis

One of the applications of conservation equation for the tumor cell density is to model *chemotaxis* phenomena. Chemotaxis is the ability of cells to respond to chemical signals by moving along the gradient of the chemical substance, either toward the higher concentration (positive chemotaxis) or away from it (negative chemotaxis). The chemical is defined as chemoattractant and chemorepellent (or chemoinhibitor), respectively.

Chemicals are not the only stimuli sensed by cells or organisms. Other stimuli include light, temperature, touch, gravity, or magnetic field. Corresponding names of phenomena induced by these stimuli are phototaxis, thermotaxis, thigmotaxis, geotaxis, and magnetotaxis. In all of those worlds the suffix *taxis* means moving towards or away from the stimulus.

Chemotaxis phenomena can be found in many species, from bacteria to multicellular organisms. They are well known in bacteria, such as *E. coli* or *S. typhimurium*, unicellular organisms, such as *Dictyostelium discoidium* and human cells, such as white blood cells (e.g., neutrophiles), sperm cells, or neurons. Chemotaxis also appears in a very important process of formation of new blood vessels by endothelial cells, termed *angiogenesis*. It is a normal process in growth and development, as well as in wound healing. However, this is also a fundamental step in the transition of tumors from a localized to an invasive state. Among many models of angiogenesis, two examples by Anderson and Chaplain [4] and Levine et al. [82] represent the continuum approach.

The chemotactic flux of cells is assumed to be given by

$$\mathbf{J}_{\text{chemotaxis}} = \chi(n, a)\nabla a, \tag{4.3}$$

where $\chi(n, a)$ is called the *chemotactic sensitivity function*, $n(\mathbf{x}, t)$ denotes density of the cells, and $a(\mathbf{x}, t)$ denotes concentration of the chemoattractant. When chemotactic sensitivity function has positive values, $\chi(n, a) > 0$, it corresponds to positive chemotaxis and a is a chemoattractor. For $\chi(n, a) < 0$ we deal with negative chemotaxis and a is a chemoinhibitor, in case that $\chi(n, a) \equiv 0$ no chemotaxis occurs.

Substituting the chemotactic flux (4.3) together with a diffusive flux into (4.1) we arrive at a basic *reaction-diffusion-chemotaxis* equation

$$\frac{\partial n}{\partial t} = \nabla \cdot (D\nabla n) - \nabla \cdot (\chi(n, a)\nabla a) + f(n), \tag{4.4}$$

where D is the diffusion coefficient of the cells. Most often, the chemotactic sensitivity function χ is assumed to be linear in the species n, hence we can write $\chi(n, a) = n\chi(a)$. Sensitivity function may assume different forms [65], for example

$$\chi(a) = \frac{\chi_0}{a}, \quad \chi(a) = \frac{\chi_0 K}{(K + a)^2}, \quad \chi_0 > 0, \quad K > 0. \tag{4.5}$$

Since the attractant $a(\mathbf{x}, t)$ is a chemical that also diffuses and is produced, we need a further equation for it. Typically,

$$\frac{\partial a}{\partial t} = D_a \nabla^2 a + g(a, n), \tag{4.6}$$

where D_a is the diffusion coefficient of a, and $g(a, n)$ is the reproduction term, which may depend on a and n. Usually it is assumed that $D_a > D$.

4.1.2 Spheroids and Tumor Cords

One of the earliest papers in the field of modeling of the spatio-temporal changes of solid tumors was published by Greenspan in 1972 [56]. He models multicellular spheroid in its steady state configuration resulting from the limitation of the oxygen diffusion towards the center of the tumor. Typically, tumors in such avascular state are limited in size to a few millimeters in diameter. Greenspan considers three layers of cells: cycling cells, viable non-proliferating cells, and necrotic cells in the core of the spheroid. Greenspan's model contains two free boundaries between different types of the cells. Disadvantage of the model is the lack of any mechanical explanation of the loss of the necrotic core volume.

Although Greenspan's model concerns the three-dimensional spheroid, time and radial distance from the core are the only variables. Spherical symmetry prevails at all times which enables to reduce the space dimension of the model to one. The reduction of dimensions is also used in modeling of *tumor cords* where cylindrical symmetry occurs. Tumor cord is a cylindrical arrangement of tumor cells, surrounded by necrosis, growing around a blood vessel of the tumor. Tumor cords are subject to the same phenomena as tumor spheroids except that now the central blood vessel is the source of nutrient for the cells resulting in opposite arrangement of cellular layers. Models of the dynamics of tumor cords were studied extensively by Bertuzzi et al. [20–22].

The structure of the cord surrounded by necrosis is as follows. Blood vessel has radius r_0 and ρ_N denote the cord radius. In the domain $r_0 < r < \rho_N$, the cell population may be subdivided into proliferating (P), quiescent (Q), and apoptotic cells (A), all surrounded by extracellular liquid. These constituents are assumed to have the same mass density, and the corresponding local volume fractions v_P, v_Q, v_A, v_E add up to one. The surface $r = \rho_N$ is the interface with the necrotic region, $\rho_N < r < B$. It is assumed that cells form a porous medium of constant porosity (that is, v_E is constant) and, irrespective of their state, they move at radial velocity $\mathbf{u} = (u(r,t),0)$. The velocity $\mathbf{v}$ of the fluid relative to the cells in the live cord is given by Darcy's law

$$v_E(\mathbf{v} - \mathbf{u}) = \kappa \nabla \tilde{p}, \tag{4.7}$$

where κ is the hydraulic conductivity and $\tilde{p}$ is the pressure of the liquid. The mass balance equations for the various cell subpopulations and for the interstitial fluid have the form

$$\frac{\partial v_P}{\partial t} + \frac{1}{r}\frac{\partial}{\partial r}(ruv_P) = \chi v_P + \gamma v_Q - \lambda v_P - \mu_P v_P, \tag{4.8}$$

$$\frac{\partial v_Q}{\partial t} + \frac{1}{r}\frac{\partial}{\partial r}(ruv_Q) = -\gamma v_Q + \lambda v_P - \mu_Q v_Q, \tag{4.9}$$

$$\frac{\partial v_A}{\partial t} + \frac{1}{r}\frac{\partial}{\partial r}(ruv_A) = \mu_P v_P + \mu_Q v_Q - \mu_A v_A, \tag{4.10}$$

$$v_E \nabla \cdot \mathbf{v} = \mu_A v_A - \chi v_P, \tag{4.11}$$

where χ is the proliferation rate, γ and λ are the rates of the transition $Q \to P$ and $P \to Q$, respectively, μ_P and μ_Q are death rates, μ_A is the volume loss rate of dead cells. The coefficients γ and λ are increasing and, respectively, decreasing functions of the oxygen concentration $\sigma(r, t)$.

Oxygen is considered a critical nutrient, although other substances (e.g., glucose or lactate) are also known to be important for the cell energy metabolism. Due to the high diffusivity of oxygen, the quasisteady-state diffusion-consumption equation

$$\nabla^2\sigma = f_P(\sigma)v_P + f_Q(\sigma)v_Q, \tag{4.12}$$

is assumed, where the functions f_P, f_Q related to oxygen consumption are of the Michaelis–Menten type.

Further transformations of Eqs. (4.7)–(4.11) lead to derivation of the PDEs for u and v and finally the equation for the pressure

$$p(r,t) = p_0(t) + \frac{v_E}{\kappa} \int_{r_0}^{r} [v(r',t) - u(r',t)]dr'. \tag{4.13}$$

4.1.3 Multiphase Theory

Another common approach in the spatio-temporal modeling of tumor growth is the multiphase theory. The theory of many interacting constituents, or *phases*, is well developed and has been successfully applied in the field of industrial applied mathematics in the 1970s and in biological sciences in the 1990s. One of the first studies devoted to multiphase modeling of tumor growth was that of Please et al. [114] in 1998, and among others one should mention at least those developed by Byrne et al. [30], Byrne and Preziosi [31], Franks et al. [50], Chaplain et al. [35], and Araujo and McElwain [6, 7].

Solid tumors are formed not only by the tumor cells, but constitute a mixture of many interacting components including normal and tumor cells, the extracellular matrix (ECM) elements, stromal cells, inflammatory cells, neural cells, immune and blood cells, and the interstitial fluid in which the components are immersed. The advantage of the multiphase theory is that it enables a tumor to be considered as multiphase material where each constituent explicitly represents a different phase. In general, each of the L phases can be described by the advection-reaction-diffusion equation

$$\frac{\partial \Phi_i}{\partial t} + \nabla \cdot (\mathbf{v}_i \Phi_i) = \nabla \cdot (D_i \nabla \Phi_i) + \Gamma_i(\Phi_1, \ldots, \Phi_L, c) - \Delta_i(\Phi_1, \ldots, \Phi_L, c) \tag{4.14}$$

where, for the i-th phase, Φ_i is the volume fraction ($\sum_{i=1}^{L} \Phi_i = 1$), $\mathbf{v}_i$ is the velocity field, D_i is the random motility or diffusion coefficient, c is the concentration of the chemical species, $\Gamma_i(\Phi_1, \ldots, \Phi_L, c)$ is the chemical- and phase-dependent production term, and $\Delta_i(\Phi_1, \ldots, \Phi_L, c)$ is the chemical- and phase-dependent degradation/death term.

Chaplain et al. [35] include in the model the effect of mechanical interactions with the surrounding tissues. They examine how incorrect sensing of compression by the cells influences cellular proliferation. In order to account for the different constituents present in the tumor the model is based on the theory of mixtures.

In addition to biomechanical effects, the model also takes into account the effect of production of the ECM and of matrix-degrading enzymes (MDEs). The ECM, on one hand side, is necessary for the cells to exist, adhere and move, and on the other it constitutes a barrier to normal cell movement. In the tissue the ECM is being constantly remodeled by the fibroblasts. It is known that tumor cells may also posses the ability to remodel the ECM; it is accomplished by production of the MDEs which degrade ECM fibers. The role of MDEs is also important at other stages of tumor growth, including invasion and metastasis; however, the manner in which they interact with inhibitors, growth factors, and tumor cells is very complex. The authors consider the cadherins (the transmembrane receptors involved in cell–cell adhesion) to be responsible for growth inhibition in case of normal cells and loss of contact responsiveness in case of tumor cells. The loss is considered to be caused by deregulation of the signalling pathway initiated by the mechanical contact between cells.

Chaplain et al. distinguish the following model variables: n, a, m_n, m_a are the volume fractions occupied by normal cells, tumor cells, host ECM, and ECM produced by tumor cells, respectively, and c is the concentration of MDEs. The system of mass balance equations for all the constituents of the model can be written as

$$
\begin{cases}
\dfrac{\partial n}{\partial t} + \nabla \cdot (n\mathbf{v}_n) = \gamma_n H_\sigma(\psi - \psi_n)n - \delta_n n\,, \\[2ex]
\dfrac{\partial a}{\partial t} + \nabla \cdot (a\mathbf{v}_a) = \gamma_a H_\sigma(\psi - \psi_a)a - \delta_a a\,, \\[2ex]
\dfrac{\partial m_n}{\partial t} = \mu_n(\Sigma(\psi))n - \nu c m_n\,, \\[2ex]
\dfrac{\partial m_a}{\partial t} = \mu_a(\Sigma(\psi))a - \nu c m_a\,, \\[2ex]
\dfrac{\partial c}{\partial t} = D\nabla^2 c + \pi_n(\Sigma(\psi))n + \pi_a(\Sigma(\psi))a - \dfrac{c}{\tau}\,,
\end{cases}
\tag{4.15}
$$

where γ_n and γ_a denote reproduction rates of normal and tumor cells, respectively; H_σ denotes the switch function (e.g., Heaviside step function, monotonic mollifier); ψ denotes the overall volume fraction; ψ_n and ψ_a denote the threshold values below which normal and tumor cells can replicate, respectively; δ_n and δ_a denote the death rates for normal and tumor cells, respectively; μ_n and μ_a denote the production rates of ECM by normal and tumor cells, respectively; ν denotes the degradation constant of ECM; D denotes the diffusion constant; π_n and π_a denote the MDE production rates, respectively, by normal and tumor cells; and τ is MDE decay constant. The relation between stress and overall volume fraction ψ is expressed by

$$
\Sigma(\psi) = E(1 - \psi_0)\left(\frac{\psi - \psi_0}{1 - \psi}\right)_+,
\tag{4.16}
$$

where ψ_0 denotes stress-free volume ratio and corresponds to the confluent volume ratio (confluency refers to the number of adherent cells in a culture dish covering entire surface); $(f)_+$ denotes the positive part of f; and E is the value of the derivative in $\psi = \psi_0$. Other forms of stress-volume ratio relations can be found in [3, 30].

Cells are regarded a granular material in a porous medium; thus following Ambrosi and Preziosi [3], one can write the momentum equation for the cells

$$\rho n \left(\frac{\partial \mathbf{v}_n}{\partial t} + \mathbf{v}_n \cdot \nabla \mathbf{v}_n \right) = \nabla \cdot \mathbf{T}_n + \mathbf{m}_n \,, \tag{4.17}$$

where $\mathbf{T}_n$ denotes the stress tensor and $\mathbf{m}_n$ is the interaction force. Neglecting inertia and introducing the usual porous media assumptions together with the assumption that normal and abnormal cells behave in the same way under compression leads to

$$\mathbf{v}_n = \mathbf{v}_a = -K \nabla \Sigma(\psi) \,, \tag{4.18}$$

where K is related to permeability of the medium.

The authors show how even a small loss of contact inhibition to growth and of compression responsiveness can lead to a clonal advantage which may in turn lead to hyperplasia and tumor growth. Among other results the most interesting is the one related to the estimate of the velocity of growth of hyperplasia which is proportional to $\sqrt{\psi_a - \psi_n}$.

Preziosi and Tosin [116] further develop the topic of mechanical interaction between cells and ECM. They introduce a general framework which allows to calculate velocities of different constituents, for example for normal and tumor cells, ECM and extracellular fluid. The example of relation between velocities of normal cells and ECM takes into account their viscoplastic interactions

$$\mathbf{v}_n - \mathbf{v}_m = \left(\frac{n}{\psi} - \frac{\sigma_{nm}}{|\nabla \cdot (\psi \mathbf{T}_\psi)|} \right)_+ \mathbf{K}_{nm} \nabla \cdot (\psi \mathbf{T}_\psi) \,, \tag{4.19}$$

where σ_{nm} denotes the threshold value for the force causing the detachment; $\mathbf{T}_\psi$ denotes the excess stress tensor; and $\mathbf{K}_{nm}$ is a positive definite matrix. A particular example of the cell momentum equation (4.19) is the case of rigid ECM and stress tensor $\mathbf{T}_\psi = -\Sigma(\psi)\mathbf{I}$. It is analogous to (4.18) and takes the form

$$\mathbf{v}_n = - \left(\frac{n}{\psi} - \frac{\sigma_{nm}}{\nabla(\psi \Sigma(\psi))} \right)_+ \mathbf{K}_{nm} \nabla(\psi \Sigma(\psi)) \,, \tag{4.20}$$

where Σ function has been defined in (4.16).

4.1.4 Modeling Desmoplastic Tumor

Publication by Psiuk-Maksymowicz [117] develops the approach proposed by Chaplain [35]. The nonlinear model developed by Psiuk-Maksymowicz concerns the growth of a desmoplastic tumor, i.e. a tumor which is enriched in fibrous connective tissue. However, unlike Preziosi and Tosin [117] focuses on development of cellular growth and loss of dependence on the ECM. Thus, the process of inducing additional apoptosis due to an inadequate number of cell–ECM bonds (called anoikis) is taken into account in the model. Another mechanism incorporated in the model relies on the stimulatory effect on cell proliferation by the same cell–ECM bonds.

Reduced model consists of three PDEs and has the form

$$
\begin{cases}
\dfrac{\partial n}{\partial t} = \overbrace{\nabla \cdot \left[nK\, \Sigma'(\psi)\nabla\psi \right]}^{\text{response to compression}} + \overbrace{\gamma_n(m)F_{\sigma_n}(\psi_n - \psi)n}^{\text{growth}} - \overbrace{\delta_n n}^{\text{apoptosis}} - \overbrace{\delta'_n F_{\sigma_n}(m_n - m)n}^{\text{anoikis}}, \\[2.2ex]
\dfrac{\partial a}{\partial t} = \overbrace{\nabla \cdot \left[aK\, \Sigma'(\psi)\nabla\psi \right]}^{\text{response to compression}} + \overbrace{\gamma_a(m)F_{\sigma_a}(\psi_a - \psi)a}^{\text{growth}} - \overbrace{\delta_a a}^{\text{apoptosis}} - \overbrace{\delta'_a F_{\sigma_a}(m_a - m)a}^{\text{anoikis}}, \\[2.2ex]
\dfrac{\partial m}{\partial t} = \overbrace{(\mu_n n + \mu_a a)F_{\sigma_m}(\psi_m - \psi)}^{\text{production}} - \overbrace{(\tilde{\pi}_n n + \tilde{\pi}_a a)m}^{\text{degradation}}.
\end{cases}
$$

$$(4.21)$$

where γ_n and γ_a denote proliferation rates of normal and tumor cells, respectively; and δ'_n and δ'_a denote degradation rates due to anoikis in normal and tumor cells, respectively. In addition to, e.g., the standard equilibrium analysis of the model the author presents simulations for the nonhomogeneous environment. The simulation results reflect the different cases of regeneration of the healthy tissue and generation of the tumor tissue. Observed asymmetry in the shape of the tumor is due to either the shape of the space domain or the spatial heterogeneity of the ECM.

4.1.5 Neovascularization Model

Another example of the application of the mixture theory is presented by Wise et al. [163]. Their nonlinear tumor growth and neovascularization model use the diffusive interface approach. The model is well-posed and consists of fourth-order nonlinear advection-reaction-diffusion equations (of Cahn–Hilliard-type) for the cell species coupled with reaction-diffusion equations for the substrate components. Understanding the morphological stability of a cancerous tumor may be important for controlling its spread to surrounding tissue. If cell–cell adhesion is uniformly high, it is expected that the resulting morphology will be compact. It is known that a compact solid tumor grows to a size limited by a diffusion of nutrient and growth factors, after which it will have to co-opt existing vasculature, or acquire a new one through angiogenesis in order to grow further [33, 56]. The model is capable of describing the dependence of cell–cell and cell–matrix adhesion on cell phenotype and genotype as well as on the local microenvironmental condition such as oxygen level. The diffusive interface model describes the dynamics of multispecies tumor growth. This approach eliminates the need to enforce complicated boundary conditions across the tumor/host interface what would have to be satisfied if the interface was assumed to be sharp. Another advantage of this method is no need to explicitly track the interface between the medium.

Approach of Wise and co-authors is based on energy variation to derive the interface equations. The system energy accounts for all the processes modeled; here, the focus is on adhesion. The diffusive interface model accounts for hydrostatic pressure, and the cell velocity, found through a generalized Darcy's law. However the modeling framework is sufficiently general to account for elastic, poroelastic, and viscoelastic effects by incorporating the relevant energies in the system energy. To solve the model numerically the authors develop an adaptive method, which solves the nonlinear equations at the implicit time step using a nonlinear multigrid finite difference method. To efficiently resolve the multiple spatial scales, an adaptive block-structured Cartesian mesh is used (see Wise et al. [162] for the details of the methodology).

Wise et al. formulate the multispecies tumor model that accounts for mechanical interactions among the different species. Primary model variables are the volume fractions of water, tumor, and host cell species, $\phi_0, \ldots, \phi_N$; and the corresponding component velocities $\mathbf{u}_0, \ldots, \mathbf{u}_N$. The mixture of $(N + 1)$-species is assumed to be saturated and thus $\sum_{i=0}^{N} \phi_i = 1$. The volume fractions obey the mass conservation equations

$$\rho_i \left(\frac{\partial \phi_i}{\partial t} + \nabla \cdot (\mathbf{u}_i \phi_i) \right) = -\nabla \cdot \mathbf{J}_i + S_i, \tag{4.22}$$

where $\mathbf{J}_i$ are fluxes and S_i are source terms that account for the intercomponent mass exchange as well as for the gains due to proliferation of the cells and the losses due to cell death. Following the approach of continuum thermodynamics, the authors derive the Helmholtz free energy of component interactions. The energy of the ith component interactions (e.g., adhesion) is represented by

$$E_i = \int \left(F_i(\phi_0, \ldots, \phi_N) + \sum_{j=0}^{N} \frac{\bar{\varepsilon}_{ij}^2}{2} |\nabla \phi_j|^2 \right) dx, \tag{4.23}$$

where the first term models the bulk energy of the component due to local interactions while the second term models longer range interactions among the components; $\bar{\varepsilon}_{ij}^2$ are positive constants, such that $\bar{\varepsilon}_{ij}^2$ has units of energy per unit length. Total energy of the system is given by $E = \sum_{i=0}^{N} E_i$. Thermodynamically consistent fluxes may take form of the generalized Fick's law

$$\mathbf{J}_i = -\bar{M}_i \nabla \left(\frac{\delta E}{\delta \phi_i} - \frac{\delta E}{\delta \phi_N} \right), \quad 1 \le i < N - 1, \tag{4.24}$$

where $\bar{M}_i > 0$ is a mobility, and $\delta E / \delta \phi_i$ are variational derivatives of the total energy E given by

$$\frac{\delta E}{\delta \phi_i} = \sum_{j=0}^{N} \left(\frac{\partial F_i}{\partial \phi_i} - \nabla \cdot (\bar{\varepsilon}_{ji}^2 \nabla \phi_i) \right), \quad 1 \le i < N, \tag{4.25}$$

The authors treat the tumor as viscous, inertialess fluid. The derived generalized Darcy laws for the velocities of the components are given by equations for u_0 i u_j

$$\mathbf{u}_0 = -\bar{k}_0 \nabla \left(\frac{\delta E}{\delta \phi_0} + q \right), \tag{4.26}$$

$$\mathbf{u}_j = -\bar{k} \left(\nabla p - \sum_{i=1}^{N} \frac{\delta E}{\delta \phi_i} \nabla \phi_i \right)$$

$$- \bar{k}_j \nabla \left(\frac{\delta E}{\delta \phi_j} - \frac{1}{\bar{\phi}_s} \sum_{i=1}^{N} \phi_i \frac{\delta E}{\delta \phi_i} + \frac{p}{\bar{\phi}_s} \right), \quad j \ge 1, \tag{4.27}$$

where q is water pressure, p is solid pressure, and $\bar{k}_0$, $\bar{k}$, and $\bar{k}_j$ are positive definite motility matrices.

The authors demonstrated analytically and numerically that when the thickness of the diffuse interface tends to zero, the system reduces to a classical sharp interface model. Simulations of the avascular tumor growth in two and three dimensions confirmed that it was feasible to compute tumor shapes with complex morphologies.

4.1.6 Modeling the Tumor Microenvironment Impact

Macklin and Lowengrub [91] investigate the tumor morphological responses depending upon the tumor microenvironment. They extend previously developed tumor growth model by allowing variability in nutrient availability and the response to proliferation-induced mechanical pressure (which models hydrostatic stress) in the tissue surrounding the tumor. Tumor volume is comprised of viable cells and necrotic cells, and those two regions are assumed to be separable. In order to perform simulations they use the level-set method [136].

General form of the multiphase equation (4.14) also includes a term corresponding to the random cell migration. The simplest version of such model is a two-phase model [156] for live and dead cells

$$\frac{\partial n}{\partial t} + \nabla \cdot (n\mathbf{v}_n) = \nabla \cdot (D\nabla n) + \Gamma(n, C_i) - \Delta(n, C_i),$$

$$\frac{\partial m}{\partial t} + \nabla \cdot (m\mathbf{v}_m) = \Delta(n, C_i), \tag{4.28}$$

where dead cells do not exhibit random motion; n and m denote volume fraction of live and dead cells, respectively; D denotes random motility; and C_i is the concentration of the chemical species (usually oxygen or a generic nutrient). It is common to assume that the convective movement of both cell types is the same, so that $\mathbf{v}_n = \mathbf{v}_m = \mathbf{v}$. Then, combining (4.28) with the volume conservation relation, we obtain an expression for cell velocity, and without a need to analyze force balances.

Usually the diffusion coefficient D is considered constant, but occasionally (4.28) is modified to include nonlinear diffusion of tumor cells [137] by setting $D = D(n, C_i)$. In the model by Sherratt [137] D is considered a decreasing function of fibrous ECM and is included in the model to represent the reduction of cell motility at high matrix densities. The model illustrates well the phenomenon of tumor encapsulation. It exhibits traveling wave solutions in which a pulse of ECM, corresponding to a capsule, moves in parallel with the advancing front of the tumor.

The general multiphase model (4.14) needs to be supplemented by equations describing the velocity of each phase. In mechanochemical models velocities are derived from the momentum balance equations. The simplest approximation assumes that the velocity takes form $\mathbf{v} = -\mu\nabla p$, where μ is a positive constant describing the viscous-like properties of tumor cells. It has analogous meaning to the Darcy's law treating the tumor tissue as a "fluid-like" structure.

4.1.7 Other Models

Another approach is presented by Byrne et al. [31] in a generic two-phase model for avascular tumor growth using the theory of mixtures. The authors assume that the cell phase can be modeled as a viscous fluid, while the water phase can be modeled as an inviscid fluid. Mass balances for the cell and water phases are expressed by typical equation of type (4.14), where random motion of those two constituents (resulting in diffusion term) is neglected. By assuming that inertial effects are negligible and no external forces act on the system, the momentum conservation laws reduce to force balances. The momentum balances for the cells, the liquid, and the mixture as a whole may be written (using $\mathbf{F}_{cw} = -\mathbf{F}_{wc}$)

$$
\begin{aligned}
0 &= \nabla \cdot (\alpha \boldsymbol{\sigma}_c) + \mathbf{F}_{cw}, \\
0 &= \nabla \cdot (\beta \boldsymbol{\sigma}_w) + \mathbf{F}_{wc}, \\
0 &= \nabla \cdot (\alpha \boldsymbol{\sigma}_c + (1 - \alpha) \boldsymbol{\sigma}_w),
\end{aligned}
\tag{4.29}
$$

where α and β denote volume fraction of cells and water, respectively; $\boldsymbol{\sigma}_c$ and $\boldsymbol{\sigma}_w$ denote volume-fraction-averaged stress tensors; $\mathbf{F}_{cw}$ denote the force exerted on the cells by water, $\mathbf{F}_{wc}$ the force exerted by the cells on water, and $\nabla \cdot (\alpha \boldsymbol{\sigma}_c)$ is the force exerted on the cells by themselves. The closure of the model occurs by choosing appropriate constitutive laws for $\boldsymbol{\sigma}_c$, and $\boldsymbol{\sigma}_w$, and $\mathbf{F}_{cw}$. The cell and water phases are treated as viscous and inviscid fluids, hence

$$
\boldsymbol{\sigma}_c = -p_c \mathbf{I} + \mu_c (\nabla \mathbf{v}_c + \nabla \mathbf{v}_c^T) + \lambda_c (\nabla \cdot \mathbf{v}_c) \mathbf{I} \quad \text{and} \quad \boldsymbol{\sigma}_w = -p_w \mathbf{I}.
\tag{4.30}
$$

In (4.30), μ_c and λ_c are the shear and bulk viscosity coefficients of the cell phase, and $\mathbf{I}$ is the identity tensor. Additionally, it is assumed that $\mathbf{F}_{cw}$ consists of contributions due to pressure and drag, hence

$$
\mathbf{F}_{wc} - p \nabla (1 - \alpha) - k(\alpha)(\mathbf{v}_w - \mathbf{v}_c),
\tag{4.31}
$$

where $k(\alpha)$ is the drag coefficient. The authors show how the model recovers less general models when asymptotic limits are considered. Unstable behavior was observed for a particular setting of the model. It is also indicated that the approach proposed is more physically based than modeling frameworks used previously to study solid tumor growth.

Apart from the multiphase models a separate category is constituted by models in which spatial dependencies of the cells are expressed by the diffusion (frequently nonlinear diffusion), such as in the model by Chaplain [33]. He presents a mathematical description of both diffusion-limited avascular and vascular tumor development. The mathematical model of the growth of a solid tumor in initial avascular stage reflects the growth of a diffusion-limited multicell spheroid. Size limitations of the multicell spheroid are associated with the chemical inhibition of mitosis. The main assumption of the model is that a growth inhibitory factor (GIF) is produced within the spheroid in a spatially dependent manner. Mitosis is assumed to be controlled by a discontinuous switch-like mechanism, such that if the concentration of the GIF is less than a threshold level θ, in any region within the tissue, mitosis occurs in this region, whereas if the concentration is greater than θ, mitosis is completely inhibited. The differential equation describing the diffusion, production, and degradation of the GIF within a spheroid can be written as

$$
\frac{\partial C}{\partial t} = \nabla \cdot (D(\mathbf{r}) \nabla C) + f(C) + \lambda S(\mathbf{r}), \quad \mathbf{r} \in \Omega,
\tag{4.32}
$$

where $C = C(\mathbf{r}, t)$ is the concentration of GIF within the spheroid occupying the region $\Omega \in R^3$, $f(C)$ is the depletion term, λ is the inhibitor production rate, and $S(\mathbf{r})$ is the source function. Chaplain examines the effect of spatial nonuniformity not on the production term but on the diffusion coefficient. Assuming radial symmetry and specific forms of the functions the model reduces to

$$\frac{\partial C}{\partial t} = \frac{1}{r^2} \frac{\partial}{\partial r} \left(r^2 D(r) \frac{\partial C}{\partial r} \right) - \gamma C + \lambda, \quad r \leq R, \tag{4.33}$$

$$\frac{\partial C}{\partial r} = 0, \quad r = 0, \tag{4.34}$$

$$D(r) \frac{\partial C}{\partial r} + PC = 0, \quad r = R, \tag{4.35}$$

where P is the permeability of the tissue surface, and the diffusion coefficient $D(r)$ is now assumed to be spatially dependent. Both monotonically increasing function of $D(r) = D(0.8 + 0.2r^2)$ and monotonically decreasing function $D(r) = D(1.0 - 0.2r^2)$ were investigated. Both examples show good agreement with the experimental observations.

Another example of the model with the nonlinear diffusion is a mathematical model to simulate brain tumor growth proposed by Swanson [147]. However, in this study the diffusion coefficient is not represented by a monotonous function but by a discontinuous function with two different values depending on the spatial location in the white or gray matter. Model describing dynamics of glioma cell density is represented by

$$\frac{\partial n}{\partial t} = \nabla \cdot (D(\mathbf{x}) \nabla n) + \alpha n, \tag{4.36}$$

where α denotes the proliferation rate, and the diffusion coefficient $D(\mathbf{x}) = D_G$ is a constant, for $\mathbf{x}$ in the gray matter $D(\mathbf{x}) = D_W$ is another constant, for $\mathbf{x}$ in the white matter such that $D_W > D_G$. Estimates of the ratio of the diffusion coefficients in gray matter and in white matter have ranged from 2 to 100-fold. The author emphasizes that it is possible to use a relatively simple model in order to model invasiveness of the gliomas which results from the nature of the disease. The model includes heterogeneous brain tissue with different motilities of glioma cells in the grey and white matter on a geometrically complex brain domain, including sulcal boundaries, with a resolution of 1 mm^3 voxels. Swanson et al. continue with models of polyclonal gliomas following chemotherapy or surgical resection which will be described in subsequent subsection. Interesting conclusion is that the velocity of the glioma expansion is linear with time and varies about tenfold, from about 4 mm/year for low-grade gliomas to about 3 mm/month for high-grade ones. An interesting extension of model with spatially heterogeneous diffusion coefficient was introduced by Painter and Hillen [109]. Their model implements anisotropic type of diffusion and utilitizes data obtained from diffusion tensor imaging.

4.2   Models of Angiogenesis

Angiogenesis is the process of new capillaries forming from existing vessels. It can be observed in the human body under various conditions. In a physiological context, it mainly takes place during embryogenesis and fetal development. Angiogenesis also is observed under pathological conditions, such as wound healing, thrombosis, and tumor growth. Angiogenesis is very important because it enables the transition of the tumor from avascular to vascular stage. The maximum size a tumor can reach without relying on vasculature for nutrient supply is about $\sim 1\,mm^3$. In the stage of avascular growth, nutrients reach the tumor by sole diffusion through the surrounding tissue. If the tumor grows beyond the size of $\sim 1\,mm^3$, cells in the core of the tumor cannot obtain enough oxygen to survive, become hypoxic and eventually starve, forming a necrotic region at the core of the tumor. Tumors can remain in the avascular state for a long time. However, one of the many responses of tumor cells to hypoxia and glucose deprivation is secretion of angiogenic growth factors that are responsible for initiating sprouting angiogenesis. Several growth factors are involved in the process of angiogenesis. Vascular endothelial growth factors (VEGF) have been identified to be one of the main driving forces [64].

Milde et al. [100] present a three-dimensional model of sprouting angiogenesis that explicitly considers the effect of the ECM and of the soluble as well as matrix-bound growth factors on capillary growth. The computational model relies on a hybrid particle-mesh representation of the blood vessels and it includes an implicit representation of the vasculature that can accommodate detailed descriptions of nutrient transport. The model combines a continuum approximation of VEGF, matrix metalloproteinases (MMPs), fibronectin, and endothelial stalk cell density with a discrete, agent-based particle representation for the tip cells. The particle representation of individual tip cells enables a grid-independent migration. Milde et al. employ particle-mesh interpolation methods to derive an implicit three-dimensional level-set representation of vessels. In most mathematical models of angiogenesis the chemotactic response of endothelial cells (ECs) to soluble VEGF gradient is considered the key driving force. The authors took into consideration the influence of the matrix-bound VEGF isoforms on the branching behavior of ECs. The simulations demonstrate that the structure and density of the ECM has a direct effect on the morphology, expansion speed, and number of branches observed in computationally grown vessel networks. Simulations of angiogenesis in the presence of matrix-bound VEGF isoforms demonstrate an increase in the number of observed branches which is consistent with the findings known from the literature.

Another example of the tumor-induced angiogenesis is considered in work by Macklin et al. [92]. Their work concerns the broader topic of multiscale modeling of vascular tumor growth. They use continuum approach to model all contained phenomenon. The model of angiogenesis accounts for blood flow through the vascular network, non-Newtonian effects, and vascular network remodeling, due to wall shear stress and mechanical stresses generated by the growing tumor. The invasion and angiogenesis models are coupled through the tumor angiogenic

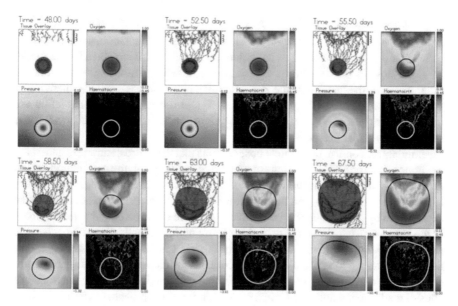

Fig. 4.1 Tumor-induced angiogenesis and vascular tumor growth. After providing a direct source of oxygen the tumor develops in a rapid way. Figure reprinted with the courtesy of Springer-Verlag [92]

factors that are released by the tumor cells, and through the nutrient extravasated from the neovascular network. As the blood flows through the neovascular network, nutrients (e.g., oxygen) are extravasated and diffuse through the ECM triggering further growth of the tumor (see Fig. 4.1). In addition, the extravasation is mediated by the hydrostatic stress generated by the growing tumor and the hydrostatic stress also affects vascular remodeling by restricting the radii of the vessels. The vascular network and tumor progression are also coupled through the ECM as both the tumor cells and the ECs upregulate matrix degrading proteolytic enzymes which cause localized degradation of the ECM which in turn affects haptotactic migration.

The endothelial cell migration and capillary sprout formation are a result of cell random motility, chemotaxis in response to tumor angiogenic factors released by the tumor (e.g., VEGF), and haptotaxis in response to ECM gradients. If n denotes the non-dimensional endothelial cell density per unit area, then the non-dimensional equation describing EC conservation is given by

$$\frac{\partial n}{\partial t} = \nabla \cdot (D\nabla n) + \nabla \cdot (\chi_{\text{sprout}}^{T}(T)n\nabla T) - \nabla \cdot (\chi_{\text{sprout}}^{E}n\nabla E) . \tag{4.37}$$

The diffusion coefficient D is assumed to be constant, and the chemotactic and haptotactic migration are characterized by the function $\chi_{\text{sprout}}^{T} = \overline{\chi}_{\text{sprout}}^{T}/(1 + \delta \cdot T)$, which reflects the decrease in chemotactic sensitivity with increased concentration T of the tumor angiogenic factor, with $\chi_{\text{sprout}}^{E} = \overline{\chi}_{\text{sprout}}^{E}$ assumed constant. Haptotactic migration is dependent on the ECM concentration, E. The coefficients D, $\overline{\chi}^T_{\text{sprout}}$, and $\overline{\chi}^E_{\text{sprout}}$ characterize the non-dimensional random, chemotactic, and haptotactic cell migration, respectively. The angiogenesis model is supplemented by the model of blood flow in the capillary network and remodeling of the entire network. In order to model the flow rates, the Poiseuille-like expression is applied, where blood viscosity is a function of capillary radius. Vessel adaptation and remodeling are modeled by means of changes in a vessel radius due to different stimuli (wall shear stress, intravascular pressure, etc.).

Simulations confirm the clinical observations that the hydrostatic stress generated by tumor cell proliferation shuts down large portions of the vascular network dramatically affecting the flow, the subsequent network remodeling, the delivery of nutrients to the tumor, and the subsequent tumor progression. In addition, ECM degradation by tumor cells is seen to have a dramatic effect on both the development of the vascular network and the growth response of the tumor. In particular, when the ECM degradation is significant, the newly formed vessels tend to surround, rather than penetrate, the tumor and are thus become less effective in delivering nutrients.

The effects on tumor growth of blood flow through a vascular network are considered by many authors. Unlike in the continuum approach, Owen et al. [108] and Welter et al. [159] use cellular automaton (CA) tumor growth models coupled with network models for the vasculature. Because of the computational cost of simulating cell growth using CA, these studies are limited to small scales. Owen et al. [108] develop a multiscale model that combines blood flow, angiogenesis, vascular remodeling, and the subcellular and tissue scale dynamics of multiple cell populations. They show that vessel pruning, due to low wall shear stress, is highly sensitive to the pressure drop across a vascular network and network remodeling is best achieved via an appropriate balance between pruning and angiogenesis. The phenomenon of formation of new blood vessels is a highly complex process which involves degradation of the ECM, EC migration and proliferation, loop formation (anastomosis) by capillary sprouts, vessel maturation, and blood flow. Further pruning and remodeling of the vascular network may be stimulated by tissue-derived signalling molecules and blood flow conditions (e.g., wall-shear stress and pressure).

The deterministic part of the model describes the cellular layer, subcellular layer, and diffusible layer (for oxygen and VEGF), while a stochastic process describes angiogenesis. In each time step Δt, at each lattice site i occupied by a vessel, new sprouts are formed with probability P_{sprout} that depends on the local concentration of VEGF via

$$P_{\text{sprout}} = \Delta t \, \frac{P^{\max}_{\text{sprout}} V}{V_{\text{sprout}} + V} , \qquad (4.38)$$

where $P^{\max}_{\text{sprout}}$ denotes maximum probability per unit time and V denotes VEGF concentration. Lateral inhibition via Delta-Notch signalling is thought to regulate the adjacent ECs sprouting. The tip of each sprout performs a random walk, biased towards regions of high VEGF concentration. The probability of moving from i to j in time Δt is defined as

$$P_{ij} = \frac{\Delta t D}{d_{ij}^2 \Delta x^2} \frac{N_m - N_j}{\sum_{k \in \Omega_i} (N_m - N_k) + N_m - N_i + N_m M} \left(1 + \gamma \frac{V_j - V_i}{d_{ij} \Delta x}\right) \quad \text{for} \quad i \neq j,$$

(4.39)

where j index denote adjacent sites to site i located in the closest proximity, N_m is the carrying capacity for movement of the cell type attempting to move, N_i is the number of cells, V_i is the VEGF at site i, γ is the chemotactic sensitivity, Ω_i is the set of sites in a neighborhood of i, d_{ij} is the distance in units of Δx between sites i and j, and D is the maximum cell mobility in the absence of chemotaxis. The way the vascular network undergoes structural adaptation (i.e., changes in their radii and hematocrit) in response to a variety of stimuli in Owen's model is similar to that presented in Macklin et al. [92]. Hematocrit is the volume percentage of red blood cells in blood; blood viscosity depends on both vessel radius and hematocrit. In addition to determining vessel radii Owen et al. compute blood flow rate, pressure drop, and hematocrit distribution in each vessel. They found that vessel pruning is principally determined by the pressure drop across the vascular network. Additionally Owen et al. found that the initial vasculature can affect the final vascular density, particularly when the pressure drop across the network is high enough to support low levels of vessel pruning.

Welter et al. [159] have focused on the arterio-venous blood vessel network remodeling during solid tumor growth. The authors are interested in the fate of the well-organized hierarchical structure of an arterio-venous vessel network remodelled by an aggressively growing tumor, as well as the global blood flow patterns and the drug transport performance of the tumor vasculature. Welter et al. formulate a hybrid cellular automaton model that predicts that the tumor vasculature is non-hierarchical and compartmentalized into a highly vascularized tumor periphery with large vessels density and dilated vessels and a central region containing necrotic regions with a low microvascular density threaded by extremely dilated vessels (see Fig. 4.2). Drug transport along the tumor vasculature turns out to be efficient in the simulations which is not consistent with the experimental observations.

The topic of mathematical modeling of tumor growth is very broad, the readers interested more in different techniques of tumor modeling in different growth stages are redirected to three review papers. In the first, by Araujo and McElwain [5], one can find a history of models from the early decades of the twentieth century to the present time. In the review by Roose et al. [129] one can find detailed data about selected continuum and discrete models of avascular tumor growth; and finally, in the very extensive paper (containing nearly 600 references) by Lowengrub et al. [90] one can find an overview of the recent results on theoretical cancer modeling (including continuum, discrete and hybrid types of models). Recent review paper by Kim et al. [72] describes different modeling approaches applied to predict spatio-temporal distributions of drugs within the tumor tissue.

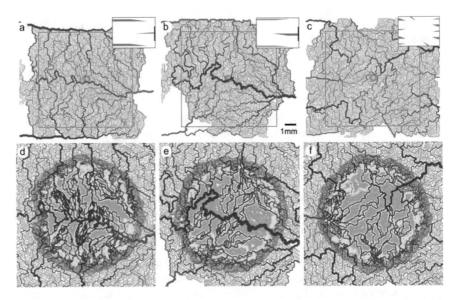

Fig. 4.2 (**a**)–(**c**) Different initial vessel networks configuration of Welter et al. [159] model. Arteries are *red* and veins *blue*. (**d**)–(**f**) Final spherical tumor with network configurations emerging from the networks above. Figure reprinted with the courtesy of Elsevier Science Ltd.

4.3 Modeling the Effects of Therapeutic Actions

Model examples presented in this section represent only models of continuous variables, CA models of the tumor growth which also take into account the therapeutic effects are placed in Sect. 4.6.

Radiotherapy is the second oldest (after surgery) type of treatment used against cancer. The most common way to model the action of ionizing radiation is to use linear-quadratic (LQ) model. The LQ formalism was originally used by Sinclair in 1966 [140]. It describes production of lethal lesions proportional to the square of the dose for the double strand breaks of the DNA and linear for the single strand breaks of the DNA. The LQ model is written as

$$S(D) = e^{-(\alpha D + \beta D^2)}, \tag{4.40}$$

where $S(D)$ is the surviving fraction of cells, D stands for radiation dose, and α and β are constant, tissue dependent model parameters. It has been found that parameters α and β correlate with the length of cell-cycle. Slow cycling tissues correspond to a smaller α/β ratio. The LQ model is still used extensively in the field, there are many extensions of this model, it may account, e.g., for separated exposure fractionation or variation in dose rate over the entire course of the radiotherapy [27]. Another mathematical measure of the radiotherapy treatment effectiveness is tumor control probability (TCP). The TCP is the probability of having no tumor cells left in the specified location after irradiation. There are different formulas for the TCP. The one derived based on a cell-cycle model is present in work by Dawson and Hillen [37].

In the multiscale model of cancer growth by Ribba et al. [124] radiosensitivity dependence on the cell cycle phase has been included. Radiation therapy is considered to act only on the proliferative cells. The authors assume that DNA damage lethal for the cells is proportional to the radiation dose according to the so-called single hit theory. Similar assumption of the linear dependence between radiation dose and therapeutic effect was used in work by Psiuk-Maksymowicz et al. [118]. In model by Ribba et al. standard fractionation therapy protocol was used in simulations, it consisted of daily 2 Gy doses applied 5 days a week, alternatively repeated for several weeks. One of the outcomes of this work is that efficacy of the treatment can be improved not necessarily by increasing the fractionation doses but by changing the treatment schedule. Such schedule should take into account features of the cell cycle of the tumor cells.

Enderling et al. [47] modeled surgical excision of the breast cancer and effectiveness of two different adjuvant treatment strategies: external beam radiotherapy (EBRT) and Targeted intra-operative radiotherapy (Targit). They use LQ model in order to predict the dose-dependent survival fractions for healthy tissue and the breast stem cells. The authors show that in contrast to conventional radiotherapy, a single-dose irradiation with Targit delivers a high dose of radiation to the tissue which surrounded the original tumor. Additionally they showed that Targit might be a better treatment strategy in terms of preventing local recurrence of the breast cancer.

Rockne et al. in the works from 2009 [126] and 2010 [127] consider brain tumor response to the EBRT. Models from both papers exploit LQ model for the radiation efficacy, whereas depending on the type of the treatment they use different formulas for the probability of cell survival $S(\mathbf{x}, t)$. Function $R(\mathbf{x}, t)$ representing effect of the radiation at location x and time t takes the form

$$R(\mathbf{x}, t) = \begin{cases} 0 & \text{for } t \notin \text{therapy} \\ 1 - S(\mathbf{x}, t) & \text{for } t \in \text{therapy} \end{cases} \tag{4.41}$$

where in the case of fractionated dose probability $S(\mathbf{x}, t)$ relies on the so-called biologically effective dose (BED):

$$S(\mathbf{x}, t) = \exp(-\alpha \text{BED}(\mathbf{x}, t)), \quad \text{BED}(\mathbf{x}, t) = nD(\mathbf{x}, t)\left(1 + \frac{D(\mathbf{x}, t)}{\alpha/\beta}\right),$$
$$\tag{4.42}$$

where n is the number of fractions, and $D(\mathbf{x}, t)$ is the fractionated dose. In the case of a single dose probability $S(\mathbf{x}, t)$ relies on effective dose E:

$$S(\mathbf{x}, t) = \exp(-E(\mathbf{x}, t)), \quad E(\mathbf{x}, t) = \alpha D(\mathbf{x}, t) + \beta[D(\mathbf{x}, t)]^2. \tag{4.43}$$

In the case of dose fractionation, different schemes were applied where fractions per day and days of treatment were subject to change. Investigations into fractionation schedules and radiation sensitivity show the highly individualized nature of dose delivery and nonlinear response to EBRT. The results suggest that the conventional treatment on a per day basis is more effective than several treatments per day. From the investigations based on the single radiation dose the authors found that the net proliferation rate is correlated with the radiation response parameter α.

Hinow et al. [66] have developed spatial model of the tumor growth where both chemotherapy and anti-angiogenic therapies are taken into account. Anti-angiogenic therapies are considered to be cytostatic in the sense that the drugs used are not toxic to the cells, but instead inhibit mechanisms necessary for the cell division. The most known anti-angiogenic drug is bevacizumab inhibiting the function of VEGF. Changes in concentration of cytotoxic drug are typically modeled by reaction-diffusion equations. The effect of the cytotoxic drug on the cells may be modeled as a linear or nonlinear effect. Depending on the drug it may be sigmoidal function of Hill-type.

Swanson et al. [147] in their model applied linear loss of tumor cells due to chemotherapy. They used piecewise constant measure of the effectiveness of the treatment. The authors have made also deliberation about the existence of drug-resistance of cell subpopulations and possibility of application. The topic of modeling the action of different drugs is connected with the pharmacodynamics and pharmacokinetics. Due to the pharmacokinetics we are able to estimate the concentration of the drug within the considered tissue.

There are also many attempts to model the action of immunotherapies. Kronik et al. [77] tried to model therapeutic vaccination against disseminated prostate cancer. Personalized models were simulated to predict changes in tumor burden and prostate-specific antigen (PSA) levels and predictions were compared to the validation set. Their model accurately predicted PSA levels over the entire measured period in 12 from 15 patients, the treatment regiment differed among the patients.

4.4   Structure in Hematopoiesis and Carcinogenesis and Its Role in Cancer Detection, Prevention, and Treatment

The unifying theme in this chapter is the role of stem cells in cancer development and the important implications that existence of normal and malignant stem cells may have for early detection, prevention, and treatment of cancer. In hematopoiesis, the key role of stem cells has been accepted for the past 100 years, because the blood production system is constantly producing massive quantities of cells, most of which quickly turn over. Therefore, its dynamics has to be rigidly regulated and even relatively small perturbations may lead to serious consequences. It seems close to a miracle that a structure which is that complicated performs its function in a stable manner.

In most other cell production systems, tissue renewal proceeds at a slower pace. A good example is the central nervous system, where until recently it was considered that no renewal was taking place in adult life. However, if a malignancy arises in a given tissue, cell renewal becomes very quick and cells acquire features such as invasiveness and motility which are absent in the normal tissue. If we assume (which seems to have become a prevailing view) that malignancies are organized around a small subpopulation of self-renewing cancer stem cells, then the hemopoietic systems become a convenient albeit simplified test-field.

In this section, we will outline the recent models taking into account the hierarchical structure of normal and malignant cell populations. We will use the hemopoietic system as a model and then extend the paradigm to other systems. We will review stochastic and deterministic models and try to determine the role of stochasticity, which has become a much debated issue. Finally, we will outline approaches taking advantage of the hierarchical structure of cancer and its stepwise progression for prevention, detection, and treatment of cancer.

4.4.1 Stochasticity or Determinism in Stem Cell Systems?

The role of stochastic events in hematopoiesis has been discussed for the past 60 years since the beginnings of experimental hematology by Till and McCulloch [74, 160] . The two opposing paradigms, deterministic hematopoiesis based on the firm regulation of peripheral blood cell populations, and stochastic hematopoiesis based on the variability observed in seeded bone marrow cells, are still awaiting a grand synthesis. This is in spite of the existence of substantial experimental findings, particularly those in the recent decade, using techniques of single-cell measurements. Disease-accompanying dynamics have been over the years variously modeled as deterministic or stochastic. Examples of stochastic phenomena observed in hematopoiesis include, but are not limited to:

- Stochastic fluctuations in the number of the hematopoetic stem cell (HSC) making self-renewal versus commitment decisions result in high variability in the magnitude of the response to infection.
- The same stochastic fluctuations may lead to depletion of the HSC compartment when facing massive infection such as in neonatal sepsis.
- Presence of variant proteins in molecular switches responding to hematopoietic growth factors such as G-CSF leads to aberrant proliferation and leukemia, again with an important chance component.
- Molecular switches under stochastic fluctuations in molecular pathways and receptor noise may become reversible, which results in reversibility and plasticity at the level of HSC and early committed cell level.

Recently, a third approach is emerging, which may be termed the molecular determinism (term coined based on ideas in [113, 141]). According to molecular determinism, stochastic variability of the proliferating bone marrow cells can be reduced to complicated series of deterministic events including molecular switches, which are multistable by nature and which trigger proliferation and/or maturation decisions. This is distinct from older proposals involving chaotic dynamics [81, 122]. Mathematical, and in particular stochastic, principles have been used to explain the balance of factors contributing to behavior of a cell population as a whole. However, new techniques for gathering data and probing biological processes at a molecule and cell level continuously provide unprecedented amounts of new information, which leads to re-examination of these models. This has led to a renewed skepticism concerning stochastic modeling as a paradigm. As argued by Snijder and Pelkmans [141], deterministic approach (or, what was called "molecular determinism") can resolve apparently stochastic phenomena with deterministic variability. They argue that cell-state parameters, such as cell size, growth rate, and cell cycle state, can be used to explain cell-to-cell variability, similarly as spatial cell population context parameters such as local cell density and location on cell colony edges. Tracing back cell-to-cell variability in time over multiple cell cycles may identify inherited, predetermining factors in cells of the same lineage. Snijder and Pelkmans [141] also advocate repeated stimulation of the same cells to help identify the presence of deterministic factors in seemingly stochastic cell-to-cell variability. Complicated dynamics leading to chaotic (and sometimes indistinguishable from stochastic) behavior has been appreciated for some time. For example, existing mathematical models of cell cycle regulation (cf. e.g., [76] and the references therein) rely on nonlinear regulatory functions to control cell population distribution. However, these models also include a very real phenomenon of uneven allocation of constituents to progeny cells, which arguably is either truly stochastic or is indistinguishable from stochastic. Moreover, the idea of "backtracking" complicated (chaotic) trajectories seems to be doubtful from mathematical viewpoint. Schroeder [134] discussed the need for long-term continuous follow-up on individual cells in order to understand the specific rules of proliferation and differentiation. This paper also touches upon issues such as influence of imaging techniques on cell behavior and difficulty with cell-tracking using existing software.

Returning to molecular determinism, a very good example of this approach seems to be the paper by Takizawa et al. [148], concerning a purely deterministic and demand-driven integrated model of regulation of early hematopoiesis. This model is very complex and it involves "view of how cytokines, chemokines, as well as conserved pathogen structures, are sensed, leading to divisional activation, proliferation, differentiation, and migration of hematopoietic stem and progenitor cells, all aimed at efficient contribution to immune responses and rapid reestablishment of hematopoietic homeostasis." Takizawa et al. [148] paper is too involved physiologically to be discussed at length here. Let us notice that it contrasts with the simpler (and stochastic) models of Ogawa [107] and Abkowitz et al. [1]. In these latter, the branching process paradigm is used at its simplest, with cells depicted as independent individuals, splitting at random and possibly interacting with a limited number of smaller entities.

Another current concept is that of non-genetic variability as a substrate for natural section, as espoused by Huang's group [29]. For example, slow fluctuations in mammalian cells are the expression of heritability (memory) of protein abundance in successive generations of normal or cancer cells [36, 139]. One example is the non-inherited form of drug resistance in cancer. Theoreticians have been suggesting this for several decades, because of similar experimental evidence. The memory of protein abundance and dynamic homeostasis, which implied slow fluctuations in individual cells, were important constituents of many of the cell cycle regulation and unequal division models [9, 152, 158]. Development of resistance to chemotherapy by gene amplification (genetic, but non-mutation driven) has been pondered by theorists equally long ago [62, 73].

Questions about the dynamics of hematopoiesis are resurfacing due to new experimental studies concerning lineage-specific growth factors, morphogens, the microenvironment, and the plasticity of stem cells [67, 78]. These new findings allow a re-examination of two long-standing questions whether hematopoiesis is stochastic or deterministic and whether it is discrete or continuous. These issues exist for other non-hematopoietic stem cell systems; however, hematopoiesis serves as the most informative and accessible mammalian tissue system to look for answers [160]. Since quantitative systems analysis based on multi-scale modeling is needed to understand the complexity and dynamics of hematopoiesis, therefore determining the correct approach to this modeling is of more than academic interest. Much work has been recently published on this topic. We will first pose three key questions and then use a simple "toy" model to explain basic ideas and problems.

Question 1. Is hematopoiesis deterministic or stochastic? Experimental data suggest stochastic processes play a role in determining cell fate of daughter cells of a stem cell [32, 57]. However it is not clear at which critical junctions stochasticity operates in lineage-specific regulation (principal examples being erythropoietin (Epo)-driven erythropoiesis and granulocyte colony-stimulating factor (GCSF)-driven granulopoiesis). Recent systemic and modeling studies of dynamics of signaling pathways in cells at various stages of hematopoiesis underscore the role of bistable (or multistable) switches, which can direct the cell towards "fates" such as differentiation in various directions, proliferation, or apoptosis [79]. These switches, as described and modeled, are essentially deterministic circuits, displaying a series of stable and unstable steady states [104]. The stable steady states correspond to distinct patterns of expression of target genes, characteristic of a given cell "fate." Small change in initial conditions at individual cell's level or in type or strength of receptor activation results in switching from one stable work regime to another [87]. Although this paradigm explains the interplay of positive and negative feedbacks in cells, it does not explain the intrinsic stochasticity, implied by both classical and more recent experiments on hematopoietic cells [1, 107]. Independently, there exists a sizeable body of evidence that eukaryotic cells may make individual decisions based on nondeterministic rules [46, 120]. The sources of intrinsic stochasticity in eukaryotic cells are related to processes in which a small number of interacting molecules may trigger a large-scale effect [84]. Stochastic effects may provide robust evolutionarily adaptive mechanisms [132]. A critical property of hematopoiesis is the ability to protect against environmental insults (e.g., infection), which may require a design incorporating stochastic dynamics.

Question 2. Do hematopoietic stem cells and their progeny constitute discrete subsets or a continuum? The general question of stem cell plasticity and, in particular, the reversibility of the hematopoietic stem cell has gained much interest due to stem cell engineering and induced pluripotent stem cells. A related question is to what extent the succession and timing and commitment and differentiation (maturation) processes in hematopoiesis can be altered or "stretched" within the bounds of normality. On an operational level, is it sufficient to model hematopoiesis in terms of discrete stages or is it necessary to include continuous maturation?

Question 3. What role is played in hematopoiesis by spatial effects? The usual approach has been to treat the process as spatially uniform both in the bone marrow and peripheral circulation. However, recent research on niches and environments in the bone marrow and the interaction of hematopoietic and mesenchymal stem cells (see [149]) has led to a realization that spatial effects and possibly interaction between spatial and stochastic effects cannot be ignored. Such interactions in mathematical models result in qualitatively new dynamics (as in Roeder's model [128], Bertolusso and Kimmel' model [18], and others [94, 102]). The reason is that spatial separation provides opportunity for small colonies of cells to fix stochastic fluctuations despite the fact the total size of the population is large.

4.4.1.1 Deterministic Models of Hematopoiesis

Configuration of feedbacks in a deterministic model of hematopoiesis. We will use as a case study the series of models devised by Arino and Kimmel [8]. Considering these models will explain the modeling paradigm, which has been later on perfected in various ways. The models are based on the following assumptions: (1) Stem cell proliferation dynamics is represented by a cell cycle model consisting of two phases: active and passive. A stem cell leaving mitosis enters the passive phase and then it may either transform into a more mature precursor cell or enter the active phase (and then divide and enter the passive phase again). It is assumed that the cell residence time in the resting phase has the exponential distribution with parameter $\alpha(t)$ (the reciprocal of the mean residence time in this phase). Such a hypothesis is consistent with the Smith-Martin model of the cell cycle. The probability of stem cell differentiation (transformation) is denoted by $d(t)$. The residence time in the active phase is equal to T. We understand that our "active phase" is $S + G_2 + M$ (see cell cycle description in Chap. 2). Our "passive phase" is assumed to be $G_0 + G_1$. (2) Regulated factors are $d(t)$ probability of stem cell differentiation and/or $\alpha(t)$ reciprocal of the mean residence time in the passive phase. (3) Each stem cell, once differentiated, produces after time H an average number of A mature (completely differentiated) cells. Quantities A and H represent all the stages of the precursor cells maturation, division, and so forth. (4) Mature cell life length is a random variable with exponential distribution with expected value $1/\beta$.

The equation for the stem cell number $N(t)$ in $G_0 + G_1$ takes the following form

$$\dot{N}(t) = -\alpha(t)N(t) + 2(1 - d(t))\alpha(t)N(t).$$

The equation for the number $R(t)$ of mature cells is

$$\dot{R}(t) = -\beta R(t) + r(t),$$

where $r(t)$ is the rate of cell flow into the mature cell compartment. Assumption (3) implies that

$$r(t) = Ad(t - H)\alpha(t - H)N(t - H)$$

so that

$$\dot{R}(t) = -\beta R(t) + Ad(t - H)\alpha(t - H)N(t - H).$$

We may also compute the number $P(t)$ of cells present at time t in the active phase of the stem cell cycle:

$$P(t) = \int_{t-T}^{t} (1 - d(\tau))\alpha(\tau)N(\tau)d\tau.$$

Equations above provide a complete description of the cell production system dynamics, if the regulated factors are specified.

Depletion and non-unique equilibria. We will make the case for the possibility of depletion of HSC and non-unique equilibria, by considering the deterministic model of erythropoietic regulation.

Model 1 The fraction $d(t)$ of differentiating stem cells is an increasing function of the number of dormant stem cells: $d(t) = g[N(t)]$. The rate $\alpha(t)$ of the outflow from the dormant stem cell compartment is a decreasing function of the number of mature cells: $\alpha(t) = h[R(t)]$. Intuitively, the mature cell number is influencing the production rate of stem cells, while the contents of the "storage" dormant compartment control the proportion of differentiating stem cells.

Model 2 In this variant, both $d(t)$ and $\alpha(t)$ depend on the mature cell number: $d(t) = g[R(t)]$, $\alpha(t) = h[R(t)]$, with $g(\cdot)$ and $h(\cdot)$ being decreasing functions. The assumption that both feedbacks here are designed to "exploit" the stem cell population causes system instability (see also [17, 85, 105, 157]).

Model 3 This is, in a sense, a reversal of Model 1. The long-range feedback controls the differentiating stem cell fraction, while the "defensive" one, the exit rate from the dormant compartment: $d(t) = g[R(t)]$, $\alpha(t) = h[N(t)]$, with $g(\cdot)$ and $h(\cdot)$ decreasing.

Model 4 This is a special case of Model 1, with $d(t) = 1/2$. In this case model equations assume the form,

$$\dot{N}(t) = -h[R(t)]N(t) + h[R(t - T)]N(t - T).$$

and

$$\dot{R}(t) = -\beta R(t) + (A/2)h[R(t - H)]N(t - H).$$

Importance of the internal feedback: Models 1, 2, and 3 have (under additional hypotheses; see the exhaustive discussion in Arino and Kimmel [8]) two equilibria, the trivial one $(N, R) = (0, 0)$ and the nontrivial one $(N, R) = (\tilde{N}, \tilde{R})$, which is a solution of nonlinear algebraic equation. Without getting into mathematical details, we can state that in Models 1 and 3, which involve autonomic internal feedbacks of the dormant HSC, the trivial equilibrium usually (i.e., for a region of parameter values) repels solutions, while the nontrivial one attracts them. Hence, the system is resistant to shocks. In Model 2, which does not include an internal feedback, the situation is reversed, given a deviation from the nontrivial equilibrium, the system decays to the trivial one. This supports the assertion that without an internal feedback, the hematopoietic system may be unstable.

Non-unique equilibria of Model 4 display an unusual behavior. Function $V(t) = N(t) + 2P(t)$ equal to the number of stem cells in the dormant phase plus twice the number of stem cells in the proliferative phase ("potential" number of HSC) stays constant along the trajectories of the system. Therefore, also at the equilibrium it will be the same value it had at time 0. Simple calculations show that in Model 4, initial conditions dictate the equilibrium value: If the system undergoes a shock such as depletion of bone marrow HSC, it will forever linger near that low value. This situation only concerns an impaired internal feedback, but it may correspond to a specific biological defect such as one caused by a defective cytokine receptor (see further on).

4.4.1.2   Leukemic Stem Cells

Among the recent comprehensive and accurate deterministic models of hematopoiesis and leukemia dynamics, it seems important to mention the models developed over recent several years by the group from Heidelberg, directed by Anna Marciniak-Czochra. The papers concern a group of topics which logically starts with analysis of dynamic feedbacks of hematopoiesis, with important distinction between self-renewal and differentiation feedbacks [54, 58, 96, 142, 143]. It is interesting to say that some ideas in these papers follow the Arino and Kimmel paper [10] discussed earlier on. Another group of papers concerns the competition of normal and leukemic (and pre-leukemic, as in the myelodysplastic syndrome) cells [155], with separate attention to modeling of efficiency of bone marrow transplant engraftment, based on clinical data [145]. Finally, the series includes papers on modeling of treatment of leukemias [144, 146], these latter also based on real-life data. Without an attempt to extend this discussion, let us notice that the papers are published in mathematical and biomedical journals, with biologists and physicians as co-authors.

4.4.1.3   Stochastic Models of Hematopoiesis

Stochastic processes (in particular the branching processes) can be used to model biological phenomena of some complexity, at cellular or subcellular levels [59]. Probabilistic population dynamics arises from the interplay of the population growth pattern with probability. Thus the classical G–W branching process defines the pattern of population growth using sums of independent and identically distributed (iid) random variables; the population evolves from generation to generation by the individuals getting iid numbers of children. This mode of proliferation is frequently referred to as "free growth" or "free reproduction." The "simple deterministic model of hematopoiesis" considered earlier on is in fact describing the expected (mean) trajectories of a multitype G–W process.

The formalism of the G–W process provides insight into one of the fundamental problems of cell populations, the extinction problem and its complement, the question of size stabilization: If a freely reproducing population does not die out, can it stabilize, or does it have to grow beyond bounds? The answer is that there are no freely reproducing populations with stable sizes. Population size stability, if it exists in the real world, is the result of forces other than individual reproduction, of the interplay between populations, and their environment. This is true for processes much more general than the G–W process. A fusion between branching and environment pressure constitutes a challenge for stochastic models (if one does not wish to resort to simulation only), the same way selection constitutes a challenge to population genetics models.

If unlimited growth models make more sense in the context of proliferation of cancer cells, then what the rate of the unlimited growth is? It can be answered within the generation counting framework of G–W type processes, but also in more general branching models. In all these frameworks, in the supercritical case, when the average number of progeny of an individual is greater than 1, the growth pattern is asymptotically exponential. The parameter of this exponential growth is the famous Malthusian parameter. In the supercritical case we can not only answer questions about the rate of growth but also questions about the asymptotic composition of non-extinct populations. What will the age distribution tend to be? What is the probability of being first-born? What is the average number of second cousins? Importantly for biological applications, many of these questions do not have natural counterparts in deterministic models of unlimited growth.

Underlying stochastic effects. Most likely mechanisms creating stochastic behavior in hematopoiesis are (1) asymmetric division of progeny cells, with resulting difference in their fates, and (2) on-off switching of differentiation status of cells accomplished by hormonal controls such as GCSF or Epo.

Asymmetric division is a possible mechanism by which randomness is inserted into stem cells decision making. Prevailing hypothesis concerning stem cell decisions was in the past that at each stem cell division, one of the progeny becomes a committed cell whereas the other remains a stem cell, in this manner providing a perfect balance between commitment and self-renewal. There exist at least two problems with this simple paradigm: first, that this does not seem satisfactory when the demand for committed cells is greater than average, and second, that it has been observed that stem cells can divide both asymmetrically and symmetrically. Observations are consistent with stochastic decisions as to which mode of division to choose. Consequences for population dynamics are different for different stochastic scenarios of asymmetric division, even if on the average these scenarios produce 50-50 committed and stem cells (see also [14, 41, 95, 103]).

An interesting discussion of symmetry and asymmetry in stem cell division has been proposed by Schroeder [133]. Discussing findings in an experimental paper by Wu et al. [164], Schroeder [133] considers a catalogue of versions of symmetric and asymmetric divisions. Symmetric division: undifferentiated hemopoietic precursor cells (HPCs) produce two undifferentiated progeny, whose later fate decisions are not linked to the parent's mitosis. Hypothetical mechanisms of asymmetric divisions in HPCs include (1) orientation of the division plane that leads to positioning of only one of the progeny close enough to localized extrinsic signals provided by a self-renewal or differentiation niche, (2) generation of two identical undifferentiated progeny, which being in close spatial contact immediately after mitosis engage in reciprocal feedback signaling, leading to differentiation of only one of them, and (3) intrinsic cell fate determinants segregate asymmetrically between daughter cells, instructing either self-renewal or differentiation of the receiving daughter. Let us notice that these distinct scenarios do not produce distinctions in deterministic models, where only averages matter, but they lead to possibly widely divergent scenarios in stochastic dynamics. As an example of mechanism (3), Wu et al. [164] found that Numb, a negative modulator of Notch signaling, which is known to asymmetrically segregate to one progeny during asymmetric division is indeed frequently enriched in one of the two emerging progeny cells. This has been accomplished by analyzing Numb localization in HPCs that had been fixed during mitosis and visualized.

Switches have been proposed to effectively translate the hormonal signals into decision about commitment or further progression. An archetypical molecular bistable switch (Gardner, Collins and Cantor genetic toggle switch; [51]) is deterministic and involves a system of two genes, the products of which are mutual cross-repressors. The system also involves two activators, which momentarily annul the action of the repressors and allow the system to switch. From mathematical point of view, the switch is a dynamical system with two stable equilibria separated by an unstable one. A more sophisticated switch, which moreover is based on confirmed molecular mechanism, is the Laslo switch [79].

However, a molecular switch may involve stochastic mechanisms, which make its action less predictable. This may mean that, if the level of fluctuations is sufficiently high, the switch is oscillating between the two stable equilibria, before or instead of being absorbed by one of them. Such behavior has been observed. In disease state, we may have to do with an aberrant switch with dynamics altered by mutation in one of the important molecular circuits.

Several transcription factors play key roles in regulating myelopoiesis and granulopoiesis. These include the ets protein PU.1 and the CCAAT enhancer binding protein α (CEBP α), and are often referred to as the "master regulators" of myeloid development [79]. Although PU.1 is sometimes considered to induce myeloid vs. lymphoid and monocyte vs. granulocyte differentiation, the data suggest that the effects of PU.1 are more complex than this. Similarly, CEBP α has been considered to direct granulocyte vs. monocyte differentiation. PU.1 and CEBP α constitute a gene regulatory network with bistable properties. Gene regulatory networks may be modified by protein abundance and post-translational modification, both of which we and others have shown to be induced by activation of cytokine receptors such as GCSFR via kinases. First, PU.1 and C/EBP α undergo serine/threonine phosphorylated triggered by GCSFR activation. Second, GCSFR activation modulates CEBP α expression, which influences PU.1 function via unidentified mechanisms. Third, GCSFR also influences Gfi-1 expression and activity.

For the granulocyte lineage, the most essential growth factor is GCSF. Its cognate receptor, GCSFR is a member of the hematopoietin cytokine receptor superfamily, which includes receptors for many of the interleukins, colony stimulating factors (e.g., Epo), cytokines (e.g., leptin), and hormones (e.g., prolactin). As a drug, recombinant human GCSF is used widely to reduce the duration of chemotherapy-induced neutropenia and mobilize into the periphery hematopoietic progenitor cells for transplant [154]. A number of clinical disorders demonstrate importance of GCSF/GCSFR (see [53, 55, 75, 130]). Mutations in the GCSFR have been found in patients with severe congenital neutropenia, myelodysplastic syndromes (MDS), and acute myeloid leukemia (AML) [11, 39].

Laslo et al. describe a regulatory network demonstrating bistability based on a feedback loop between two transcriptional repressors (Egr/Nab-2 and Gfi-1) of PU.1 and GATA-1 genes that drive a common myeloid progenitor cell toward either granulocyte or macrophage fate. As mentioned above, deterministic toggle, or bistable, switch is a circuit which has two stable equilibria, usually separated by an unstable one [51]. Stochastic toggle switches have much richer behavior [86]. Instead of a monotonous approach to the stable equilibrium, the absorbing state is reached via a "saw-like" trajectory. If the time before absorption extends over more than a single cell cycle, the cell remains uncommitted, or in one of the "intermediate" states, as, for example, in the paper by Laslo et al. [80] where existence of graded states of cells was experimentally observed and theoretically predicted (albeit using a deterministic switch). On the theoretical side, state space methods have been used by Michaels et al. [99] to find the range of dynamical behaviors exhibited by Laslo-type switch.

Jaruszewicz et al. [68] demonstrate that in a system of bistable genetic switch, the randomness characteristics control in which of the two epigenetic attractors the cell population will settle. They focus on two types of randomness: the one related to gene switching and the one related to protein dimerization. Change of relative magnitudes of these random components for one of the two competing genes introduces a large asymmetry of the protein stationary probability distribution and changes the relative probability of individual gene activation. Increase of randomness associated with a given gene can both promote and suppress activation of the gene. Each gene is repressed by an increase of gene switching randomness and activated by an increase of protein dimerization randomness. In summary, the authors demonstrated that randomness may determine the relative strength of the epigenetic attractors, which may provide a unique mode of control of cell fate decisions.

Traulsen et al. [151] concentrate on the role of hierarchy of the hematopoietic system, discussing the influence of mutations in the hematopoietic system. Although mutations can occur in any cell within hematopoiesis, both the size of the circulating clone and its lifetime depend on the location of the cell of origin in the hematopoietic hierarchy. Mutations in more primitive cells give rise to larger clones that survive for longer, taking also a longer time to appear in the circulation. On the contrary, the smaller clones caused by mutations of more differentiated precursors appear in the circulation much more rapidly after the causal mutation, but they are smaller and survive shorter. Three disease-causing mutations serve as illustrations: the BCR-ABL associated with chronic myeloid leukemia; mutations of the PIG-A gene associated with paroxysmal nocturnal hemoglobinuria; and the V617F mutation in the JAK2 gene associated with myeloproliferative diseases. Among others, evidence is presented of existence of these mutations in asymptomatic individuals, speculatively, these are mutations in more differentiated precursors. Citing from Traulsen et al. [151]: "In general, we can expect that only a mutation in a hematopoietic stem cell will give long-term disease; the same mutation taking place in a cell located more downstream may produce just a ripple in the hematopoietic ocean."

Wilson et al. [161] argue based on a combination of flow cytometry with label-retaining assays (BrdU and histone H2B-GFP) that there exists a population of dormant mouse HSCs (d-HSCs) within the lin− Sca1+ cKit+ CD150+ CD48− CD34− population. Computational modeling suggests that d-HSCs divide about every 145 days, or five times per lifetime. d-HSCs harbor the vast majority of multilineage long-term self-renewal activity. They form a reservoir of the most potent HSCs during homeostasis, and are efficiently activated to self-renew in response to bone marrow injury or G-CSF stimulation. After re-establishment of homeostasis, activated HSCs return to dormancy, suggesting that HSCs are not stochastically entering the cell cycle but they reversibly switch from dormancy to self-renewal under conditions of hematopoietic stress.

Becker et al. [13] show that Epo receptors have the ability to cope with steady-state and acute demand in the hematopoietic system. By mathematical modeling of quantitative data and experimental validation these authors showed that rapid ligand depletion and replenishment of the cell surface receptor are characteristic features of the erythropoietin receptor (EpoR). The amount of Epo-EpoR complexes and EpoR activation integrated over time corresponds linearly to ligand input.

4.4.2   Models of Mutations and Evolution of Disease

4.4.2.1   Carcinogenesis Models

Carcinogenesis modeling has had an established history of using stochastic models, beginning with the Knudson two-hit model. Successor models include the multi-hit model and eventually to the two-stage clonal expansion model of Moolgavkar [101]. With almost 1000 citations, this paper might be called one of the most influential ever mathematical models in cancer research. Concerning its application in leukemias, see, e.g., Radivoyevitch et al. [119].

We will focus mostly on models conceived in the genome-sequencing era. These models have the following features, which are a novelty due to both evolution of thinking in inflow of a large number of new variant data:

1. Mutations are identified as variants in studies in which whole exomes or even genomes are sequenced for each individuals in the study.
2. Functionality of mutations is determined in two stages: first by bioinformatics algorithms (usually based on evolutionary comparisons) and then by wet-laboratory studies of pathways influenced by these mutations.
3. Progression of carcinogenesis is based on the concept of driver and passenger mutations. Driver mutations are selected for most advantageous phenotype of cancer cells, whereas the passenger mutations are neutral by-products of carcinogenesis and serve as molecular clocks of the process.

A number of interesting models of mutations leading to cancer have recently been published (see references further on). They all explore models of proliferation, frequently using branching processes, combining them with models of driver and passenger mutations. Driver mutations are those that, although they might have arisen spontaneously, provide selective advantage for the emerging cancer prolifer-ation, particularly against the background of already existing inherited or acquired mutations. Passenger mutations are generally neutral and their accumulation may provide a molecular "clock" indicating how long it has been since the cancer cells deviated from normal cells. Tumors are initiated by the first genetic alteration that provides a relative fitness advantage. In the case of leukemias, this might represent the first alteration of an oncogene, such as a translocation between BCR (breakpoint cluster region gene) and ABL (V-abl Abelson murine leukemia viral oncogene homolog 1 gene).

Recent paper by Ley et al. [83] addressed the issue of driver mutations contribut-ing to the pathogenesis of AML, using analysis the genomes of 200 adult cases of de novo AML, either whole-genome sequencing (50 cases) or whole-exome sequencing (150 cases), along with RNA and microRNA sequencing and DNA-methylation analysis. The conclusion was that AML genomes have fewer mutations than most other adult cancers, with an average of only 13 mutations found in genes. Of these, an average of 5 were in genes that are recurrently mutated in AML. A total of 23 genes were significantly mutated, and another 237 were mutated in two or more samples. Further analysis suggested strong biologic relationships among several of the genes and categories. Further studies of this kind are likely to lead to insights into the nature of these relationships, although with very few exceptions (such as [15]) only a single time point in patient lifetime is usually available.

One of the important recent papers with this focus is the mathematical model of the relationship between accumulation of driver and passenger mutation in tumors published by Nowak and Vogelstein groups [25]. In most previous models of tumor evolution, mutations accumulate in cell populations of constant size or of variable size, but the models take into account only one or two mutations. Such models typically address certain aspects of cancer evolution, but not the whole process. In the model presented in Bozic et al. [25], it has been assumed that each new driver mutation leads to a slightly faster tumor growth rate. This model is as simple as possible, because the analytical results depend on only three parameters: the average driver mutation rate u, the average selective advantage associated with driver mutations s, and the average cell division time T. The model is based on the Galton–Watson branching process.

The hypotheses are as follows: At each time step, a cell can either divide or differentiate, senesce, or die. In the context of tumor expansion, there is no difference between differentiation, death, and senescence, because none of these processes will result in a greater number of tumor cells than present prior to that time step. It is assumed that driver mutations reduce the probability that the cell will become "stagnate," i.e., that it will differentiate, die, or senesce, although the stagnant cells are not removed from the tumor. A cell with k driver mutations has a stagnation probability $d_k = (1 - s)^k/2$. The division probability is $b_k = 1 - d_k$. The parameter s is the selective advantage provided by a driver mutation. When a cell divides, one of the daughter cells can acquire an additional driver mutation with probability u. The theory can accommodate any realistic mutation rate and the major numerical results are only weakly affected by varying the mutation rate.

We can calculate the average time between the appearances of successful cell lineages. Not all new mutants are successful, because stochastic fluctuations may lead to the extinction of a lineage. The lineage of a cell with k driver mutations survives only with a probability of approximately $1 - d_k/b_k \approx 2sk$. Assuming that $u \ll ks \ll 1$, the average time between the first successful cell with k and the first successful cell with $k + 1$ driver mutations is given by

$$\tau_k = \frac{T}{ks} \ln \frac{2ks}{u}.$$

This result is obtainable from the theory of the G–W process (by elementary means) and the derivation is found in the supplement to Bozic et al. [25]. The cumulative time to accumulate k mutations grows logarithmically with k. On the other hand, the average number of passenger mutations, present in a tumor cell after t days is proportional to t, that is, where is the rate of acquisition of neutral mutations. Combining the results for driver and passenger mutations, results in a formula for the number of passenger mutations that are expected in a tumor that has accumulated

k driver mutations

$$n = \frac{\nu}{2s} \ln \frac{4ks^2}{u^2} \ln k.$$

Here, n is the number of passengers that were present in the last cell that clonally expanded. Bozic et al. [25] demonstrate that this dependence fits empirical data on several human cancers.

4.4.2.2 Distribution of Mutational Events in Various Phases of Premalignant and Malignant Proliferation

This question has been recently addressed by Tomasetti et al. [150]. The framework is not very different from that of Bozic et al. [25]. However, the paper describes mathematically the different phases in which somatic mutations occur in a tissue giving rise to a cancer. Starting from a single fertilized egg, all tissues are created via clonal expansion (development phase). The tissue is then subjected to periodic self-renewals. During development and tissue renewal, passenger mutations occur randomly, undergo clonal expansions, and either become extinct or expand as successive passenger mutations accumulate. A driver gene mutation may initiate a tumor cell clone, which then can expand through subsequent driver mutations, eventually yielding a clinically detectable tumor mass (cancerous phase). Passenger mutations occur during this phase as well. The model makes the novel prediction, validated by empirical findings, that the number of somatic mutations in tumors of self-renewing tissues is positively correlated with the age of the patient at diagnosis. Importantly, the analysis indicates that half or more of the somatic (i.e., acquired, non-germline) mutations in tumors of self-renewing tissues occur prior to the onset of neoplasia. This is the case, among others for the chronic lymphocytic leukemia (CLL). The model also provides a novel way to estimate the in vivo tissue-specific somatic mutation rates in normal tissues directly from the sequencing data of tumors.

Stochastic models also allow modeling of eradication of leukemic stem cells while saving healthy stem cells. The paper by Sehl et al. [135] uses an impressive array of analytical and computational tools to analyze a pair of stochastic processes describing proliferation and death of healthy and cancer stem cells under chemotherapy. The question asked concerns the birth and death rates differential between these two cell types, required to eradicate the latter and preserve the former. Mutations, emergence of drug resistance, interactions of cancer and healthy cells, and other complicating factors are disregarded. Because the biological setup is simplified to the extreme, it allows effective mathematical analysis.

A review of gene copy number and loss of heterozygosity and gene mutation profiles demonstrated that relapsed AML invariably represented reemergence or evolution of a founder clone [110]. Analysis of informative paired persistent AML disease samples uncovered cases with two coexisting dominant clones of which at least one was chemotherapy sensitive and one resistant, respectively. These data support the conclusion that incomplete eradication of AML founder clones rather than stochastic emergence of fully unrelated novel clones underlies AML relapse and persistence.

As a side note, it seems surprising that quite few papers offer estimates of absolute numbers of hematopoietic stem cells and committed cells. Against this background, the paper by Peixoto et al. [112] discusses the mathematics of hematopoiesis based on stochastic hypotheses. Mathematical model that describes normal hematopoiesis across mammals as a stable steady state of a hierarchical stochastic process is also used to understand the detailed dynamics of a range of blood disorders both in humans and in animal models. The paper includes comparative numerical estimates of the numbers of cells in different compartments.

4.4.3   Prevention, Detection, and Treatment of Cancer Given Hierarchical Structure of Cancer and Its Stepwise Progression

Advances in the understanding of stem cell proliferation stimulated stochastic modeling to understand the reasons for failure or success of chemotherapy in acute leukemia. The paper by Sehl et al. [135] uses an impressive array of analytical and computational tools to analyze a pair of stochastic processes describing proliferation and death of healthy and cancer stem cells under chemotherapy. The question asked concerns the birth and death rates differential between these two cell types, required to eradicate the latter and preserve the former. Mutations, emergence of drug resistance, interactions of cancer and healthy cells, and other complicating factors are disregarded. Because the biological setup is simplified to the extreme, it allows effective mathematical analysis, at least up to a point. The approach used is quite sophisticated, for example, the distribution of the number of surviving healthy stem cells is considered at the random stopping time defined by extinction of the cancer stem cells. In addition, nontrivial applications of extreme-value theory allow obtaining asymptotic distributions of the times to extinction for cell populations started by multiple ancestors. Conclusions reached are interesting, even that it seems difficult to relate the simplified setup to biologically relevant situation. There exist rich literature concerning scheduling of leukemia treatment; one example based on the model by Roeder et at. [128] has been described mathematically in [71].

4.5   Physiologically Structured Models

The structured models connect the kinetics of cell populations with biological processes in the cell cycle. These models identify such properties as age, volume, mass, DNA content, RNA content, maturity, or other characteristics of individual cells as structure variables. Three basic types of the model include: transition probability models, size control models, and inherited property models [158]. Transition probability models are examples of quasi-probabilistic models of the cell cycle. These models assume that variability in the cycle is confined almost entirely to G_1. Cell-cycle is divided into the indeterminate A-phase of variable length and B-phase of constant length. The A-phase is part of G_1, and the B-phase is the rest of G_1, plus S, G_2, and M. Transition probability models provide good agreement with experimental data.

Size control models are examples of deterministic models of the cell cycle. These models assign fundamental control to cell size as cells progress through the cell cycle. Size control models rely on the experimental evidence that cell size does influence the timing of cell division. Many authors [19, 106, 152] claim a critical size requirement for entry into S-phase or M-phase or both.

Inherited property models of the cell cycle are based upon the assumption that cell behavior is in some part determined by cell pedigree. In these models it is assumed that the transit time through the cell cycle is determined at birth and variation of transit times results from variation of initial states of newborn cells.

A particular example of the transition probability model is the model of proliferating and quiescent cells which exhibit asynchronous exponential growth [45]. Dyson et al. [45] analyze a linear model of an age-structured cell population with proliferating and quiescent compartments. Due to linearity of the model it is applicable to experimental cell cultures, early stage tumors, and other populations before resource limitations produce the nonlinear effects of crowding. A characteristic of asynchronous exponential growth in an age-structured population is that the total population grows exponentially, but the age structure stabilizes in the sense that the fraction of cells in any age range converges to a limit value independent of the initial age distribution. Another feature of their model is that both the proliferating and quiescent subpopulations ultimately disperse over all ages independently of any initial age or proliferating-quiescent synchronization.

The age densities of proliferating and quiescent cells at time t, respectively, $p(a, t)$ and $q(a, t)$, are the model unknowns. The authors consider a system of linear partial differential equations

$$\frac{\partial p}{\partial t} + \frac{\partial p}{\partial a} = -(\mu(a) + \sigma(a))p + \tau(a)q \,,$$

$$\frac{\partial q}{\partial t} + \frac{\partial q}{\partial a} = \sigma(a)p - \tau(a)q \,, \tag{4.44}$$

$$p(0, t) = 2f \int_0^{a_1} \mu(a)p(a, t)da \,, \quad q(0, t) = 2(1-f) \int_0^{a_1} \mu(a)p(a, t)da \,,$$

$$p(a, 0) = p_0(a) \,, \quad q(a, 0) = q_0(a) \,,$$

where the function $\mu(a)$ is the age-dependent rate of division of proliferating cells, the functions $\sigma(a)$ and $\tau(a)$ are the age-dependent rates of transition between the proliferative and quiescent compartments, and f is a constant, $0 < f \le 1$. In an age-structured population the mechanism driving asynchronous exponential growth is the variability of the age of division. Dyson et al. [45] established sufficient conditions on the functions controlling division and transition between proliferating and quiescent compartments to ensure asynchronization of the evolving population.

In order to take into account physiological properties of cancer cells (e.g., mass, volume, DNA content) the McKendrick [97] PDE framework is most suitable. Physiologically structured models have been extensively studied, see, e.g., [23, 38, 43]. General model of four cell-cycle stages (G_1, S, G_2, and M) is represented by set of equations

$$\frac{\partial n_i(t, x)}{\partial t} + \frac{\partial n_i(t, x)}{\partial x} + d_i(t, x)n_i(t, x) + K_{i \to i+1}(t, x)n_i(t, x) = 0,$$

$$n_{i+1}(t, 0) = \int_0^\infty K_{i \to i+1}(t, x)n_i(t, x)dx, \tag{4.45}$$

$$n_1(t, 0) = 2 \int_0^\infty K_{I \to 1}(t, x)n_I(t, x)dx,$$

where $n_i(t, x)$ denotes the densities of cells having age x in phase i, d_i is the tumor death rate, $K_{i \to i+1}$ is the tumor cell transition kernel between phases i and $i + 1$, e.g., G_1 and S.

A way to represent cell population growth control by cytostatics in age-structured models is to send proliferating cells to a quiescent phase where they do not proliferate [23]. The rate of proliferating cells that become quiescent might be expressed as an increasing function of the cytostatic drug dose. This linear McKendrick model for proliferative compartment and a quiescent one, where the effect of erlotinib–cytotoxic drug is incorporated, is written as

$$\frac{\partial}{\partial t}p(t, x) + \frac{\partial}{\partial x}p(t, x) + (\mu + K(x))p(t, x) = 0,$$

$$p(t, 0) = 2(1 - f) \int_{\xi \geq 0} K(\xi)p(t, \xi)d\xi,$$

$$p(0, x) = p_0(x), \tag{4.46}$$

$$\frac{d}{dt}Q(t) = 2f \int_{\xi \geq 0} K(\xi)p(t, \xi)d\xi - \nu Q(t),$$

$$Q(0) = Q_0,$$

and the drug target is denoted f, rate of escape at mitosis towards the siding phase is denoted Q, with f erased by a cytostatic drug.

There are many other extensions to the age-structured models. One of them may include delay effect in order to model maturation of the cells [49], other models incorporate space in which the interactions between the tumor and its environment are investigated [28, 125]. A nonlinear model of age and maturity structured population dynamics was analyzed by Dyson et al. [44]. The authors studied existence and asymptotic behavior of solutions of the model. Their population is described by a density function $u(t, a, x)$, where t is time, a is age of individuals, and x is maturity of individuals. Cells of all maturity values are capable of division and dividing cells enter the population with cell age 0. The model has the form

$$\tfrac{\partial}{\partial t} u(t, a, x) + \tfrac{\partial}{\partial a} u(t, a, x) + x \tfrac{\partial}{\partial x} u(t, a, x)$$

$$= (\nu(a) + \mu(x) - \eta(\mathcal{F}(u(\cdot, \cdot, t)))) u(t, a, x),$$

$$t > 0, a > 0, 0 \leq x \leq 1 \tag{4.47}$$

$$u(t, 0, x) = \int_0^\infty \beta(a) u(t, a, x) da, \qquad t > 0, 0 \leq x \leq 1$$

$$u(0, a, x) = \phi(a, x), \qquad a > 0, 0 \leq x \leq 1,$$

where the coefficient x of the partial derivative $\tfrac{\partial}{\partial x} u(t, a, x)$ corresponds to the rate of maturation of individual cells. All cells mature at the same rate and the time for a cell to mature from x_1 to x_2 is $\int_{x_1}^{x_2} \tfrac{1}{x} dx$. Since x_1 may approach 0, the most immature cells require a very long time to reach the higher levels of maturity. The linear coefficients $\nu(a)$ and $\mu(x)$ account for entry or exit of individuals with dependence on age a and maturity x, respectively, and the nonlinear coefficient $\mathcal{F}(u(\cdot, \cdot, t))$ accounts for loss of individuals due to crowding (η is a scalar function and $\mathcal{F}$ is a linear functional of the density $u(\cdot, \cdot, t)$). The coefficient $\beta(a)$ accounts for the birth process of a daughter cell of age 0 from a mother cell of age a. The initial distribution of the population at time 0 is given by $\phi(a, x,)$. In order to analyze the model the nonlinearity corresponding to nonlocal crowding effects on mortality was handled by the authors as a scalar perturbation of the linear problem.

A different type of the structured model was recently introduced by Lorz et al. [89]. In this work cancer cells are assumed to be structured as a population by two real variables standing for space position and the expression level of a phenotype of resistance to cytotoxic drugs. The model explicitly takes into account the dynamics of resources and anticancer drugs as well as their interactions with the cell population under treatment. They analyze the effects of space structure and combination therapies on phenotypic heterogeneity and chemotherapeutic resistance.

Cells are structured as a population by two nonnegative real variables $x \in [0; 1]$ and $r \in [0; 1]$ standing, respectively, for the normalized expression level of a cytotoxic-resistant phenotype and for the normalized linear distance from the center of the spheroid. This implies that cells do not have to be totally sensitive or totally resistant to a given drug but may take an intermediate state. Both structure variables x and r have a well-defined biological meaning. For example, a cell resistance level can be measured by the expression level of ABC transporter genes that are known to be associated with resistance to the drug.

The population density of cancer cells is modeled by the function $n(t, r, x) \geq 0$, so that the local and global population densities at time $t \geq 0$ are computed, respectively, as

$$\rho(t, r) = \int_0^1 n(t, r, x)dx, \qquad \rho_T(t) = \int_0^1 \rho(t, r)r^2 dr .$$

The function $s(t, r) \geq 0$ identifies the concentration of resources available to cells. The densities of cytotoxic and cytostatic drugs are described, respectively, by $c_1(t, r) \geq 0$ and $c_2(t, r) \geq 0$.

The dynamics of functions n, s, c_1, and c_2 is ruled by the following set of equations:

$$\partial_t n(t, r, x) = \left[\frac{p(x)}{1 + \mu_2 c_2(t, r)} s(t, r) - d\rho(t, r) - \mu_1(x)c_1(t, r) \right] n(t, r, x),$$

$$-\alpha_s \Delta s(t, r) + \left[\gamma_s \int_0^1 p(x)n(t, r, x)dx \right] s(t, r) = 0, \qquad (4.48)$$

$$-\alpha_{c_1} \Delta c_1(t, r) + \left[\gamma_{c_1} \int_0^1 \mu_1(x)n(t, r, x)dx \right] c_1(t, r) = 0,$$

$$-\alpha_{c_2} \Delta c_2(t, r) + \left[\gamma_{c_2} \mu_2 \int_0^1 n(t, r, x)dx \right] c_2(t, r) = 0,$$

where Δ stands for the Laplacian in polar coordinates; the function $p(x)$ models the proliferation rate of cells expressing the resistance level x due to the consumption of resources; the parameter d is the average death rate of cells due to the competition for limited space; the function $\mu_1(x)$ denotes the death rate of cells due to the consumption of cytotoxic drugs; the parameters α_s, α_{c_1}, and α_{c_2} model, respectively, the diffusion constants of resources, cytotoxic drugs, and cytostatic drugs; the parameters γ_s, γ_{c_1}, and γ_{c_1} represent the decay rate of resources, cytotoxic drugs, and cytostatic drugs, respectively.

The model defines zero Neumann conditions at $r = 0$ coming from radial symmetry and Dirichlet boundary conditions at $r = 1$

$$s(t, r = 1) = S_1(t) \quad \partial_r s(t, r = 0) = 0$$
$$c_{1,2}(t, r = 1) = C_{1,2}(t) \quad \partial_r c_{1,2}(t, r = 0) = 0 .$$

The authors claim that phenotypic heterogeneity within solid tumors can be explained in terms of selection driven by the local cell environment. Another finding is that cytostatic drugs can slow down tumor progression, while cytotoxic drugs can favor the selection of highly resistant cancer clones and cause a decrease in the heterogeneity with respect to the resistance trait, which is in agreement with the Gause's competitive exclusion principle. Study of different therapeutic protocols showed that combination of both cytotoxic and cytostatic drugs is more effective than one type of the drugs. Their studies did not show unambiguously whether bang-bang infusion of cytotoxic drug and constant delivery of cytostatic drug is in favor of constant supply of cytotoxic and cytostatic drugs.

Billy et al. [24] showed how the modeling approach allows to design original optimization methods for anticancer therapeutics, in particular chronotherapeutics. They have represented the dynamics of the division cycle in proliferating cell prolif-erations by physiologically structured partial differential equations. The anticancer drug effect is represented by a blockade of the cell cycle at the main checkpoints G_1/S and G_2/M. Billy et al. mathematically represented and optimized the effects of a time-dependent blockade of cell cycle phase transitions in proliferating cell populations. In order to resolve the optimization problem they have used the projected gradient method. The authors also proposed a numerical algorithm to control eigenvalues of the differential operators underlying the cell proliferation dynamics, in tumor and in healthy cell populations. Their studies suggest infusion schedules set a maximal drug infusion flow when cancer cells get hurt by the drug while healthy cells do not.

4.6 Cellular Automata and Agent Based Approach to Treatment Planning

4.6.1 Biologically Motivated Cellular Automata

One of the tools to model complex physical phenomena through computer simula-tions are cellular automata (CA). The CA are a discrete model of time, space, and state in which the physical laws are mimicked by a series of simple rules that are easy to compute quickly and in parallel [48]. The first studies of the CA originate from the 1940s. This method is useful in solving large systems of nonlinear integro- and partial-differential equations. This is particularly true for biological phenomena. Typically, continuum models involve averaging out the spatial organization in order to obtain simplified and tractable equation. In case of CA, solution of the model is exact in space and it allows to examine spatial and temporal pattern formation of the system.

It is possible to distinguish at least three classes of CA: (1) deterministic automata; (2) lattice gas models; and (3) "solidification" models. In a deterministic model each lattice point has associated a state, which is determined in next time step solely based on the earlier states of the cell and its neighbors. In lattice gas model the system is usually driven by random events. The "solidification" models are much like lattice gas models except binding the particle state in space. This type of models has been used to model phase-transitions, angiogenic branching growth or growth of bacteria colonies under nutrient limitations. The CA play an important role in theoretical biology. They provide a simple transition from a verbal statement of the mechanisms to a formal model, and due to speed and simplicity they are good tools for testing a range of models and parameters.

Biological systems are particularly suitable for being analyzed by cellular automata. Alarcon et al. [2] apply CA to model growth of a tumor in inhomogeneous environment. The authors try to show how blood flow and red blood cell heterogeneity influence the growth of normal and cancerous cells. They first calculate the distribution of oxygen in existing vascular network, taking into consideration the features of blood flow and vascular dynamics such as structural adaptation, complex rheology, and red blood cell circulation. Subsequently, they evaluate the dynamics of a colony of normal and cancerous cells, placed in such heterogeneous environment. The cells and oxygen particles are considered elements of a cellular automaton. The two-dimensional model consists of an array of $N \times N$ automaton elements. The state of each element is defined by a state vector, which has three components: occupation, cell status, and oxygen concentration. The state vector evolves according to prescribed local rules. The authors consider a simple square lattice, i.e. each cell has four (nearest) neighbors (the so-called von Neumann neighborhood). The rules of the CA are inspired by generic features of tumor cells, such as the ability to endure very low levels of oxygen, or competition between cancer and normal cells for nutrients. The important step is to formulate the boundary value problem for the extracellular oxygen concentration. No-flux boundary conditions are imposed along the edges of the domain. At the walls of the vessels a mixed type boundary condition is imposed. By means of the CA, Alarcon et al. were able to capture generic features of the growing tumor. They found that environmental inhomogeneity, caused by non-uniform oxygen perfusion, significantly decreases the ability of cancerous cells to grow and invade healthy tissue. The weakness of the model is that blood vasculature is fixed and does not adapt to the surrounding conditions.

Kansal et al. [70] in 2000 have presented different types of CA model. They have developed a three-dimensional CA model of brain tumor growth on a Voronoi lattice. In the three spatial dimensions the Voronoi cells take the form of polyhedra. The advantage of this type of lattice is that it preserves the discrete nature of cells but removes the anisotropy introduced by the use of a regular lattice. The model uses a varying density of lattice sites (an adaptive grid lattice). This allows small tumors to be simulated with greater accuracy, while still allowing the tumor to grow to a large size. This model simulates Gompertzian growth for a tumor growing over nearly three orders of magnitude in radius (see Fig. 4.3). It also predicts the composition of the cells within a tumor and tumor dynamics in agreement with medical data. The model enables to directly simulate more complex physiological situations, e.g. coexistence of phenotypically different tumor subpopulations.

Another example of application of the CA in theoretical biology is one developed by Patel et al. [111]. They study the role of acidity in tumor growth taking advantage of the cellular automaton model. Additionally they examined the roles of host tissue vascular density and tumor metabolism in the ability of a small number of monoclonal transformed cells to develop into an invasive tumor. Patel et al. have constructed a hybrid cellular automaton model capable of simulating early tumor growth. The model incorporates normal cells, tumor cells, necrotic and empty space, and a random network of native microvessels as components of the CA state

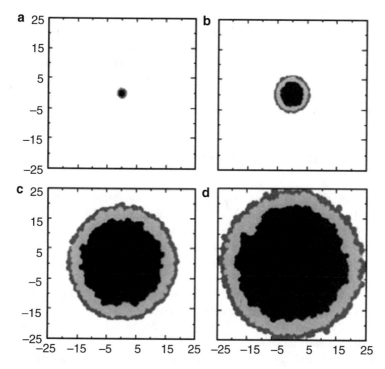

Fig. 4.3 The development of the cross-central section of a tumor in time. (**a**) corresponds to the tumor spheroid stage, (**b**) to the first detectable lesion, (**c**) to diagnosis and (**d**) to death. The dark-gray outer region is comprised of proliferating cells, the light-gray region is non-proliferative cells and the black region is necrotic cells. The scales are in millimeters. Figure reprinted with the courtesy of Academic Press [70]

vector. Microvessel automaton elements are randomly distributed throughout the simulation space. Continuous part of the model is described by partial differential equations for the glucose and H^+ ion concentrations. Occurrence of the H^+ ions results largely from the tumor's excessive anaerobic metabolism. The cells and microvessels affect the extracellular concentrations of glucose and H^+ which, in turn, affect the evolution of the automaton. The authors have shown that high H^+ ion production is favorable for tumor growth and invasion. There is minimal value of microvessel density below which both tumor and normal cells die due to excessively low pH. There also exists the level of vascular densities above which the microvessel network is highly efficient at removing acid and therefore the tumor cells lose their advantage over normal cells they have in by high local H^+ concentration. Their simulations showed variety of tumor morphologies ranging from compact tumors to those with diffuse necrosis, isolated necrosis or tumor cords.

Hatzikirou and Deutsch [63] in their 2008 work focused on the analysis of cell migration strategies involving complex feedback mechanisms between the cells and their microenvironment. This interplay is essential for central biological processes as embryonic morphogenesis, wound healing, immune reactions, and tumor growth.

They introduced lattice-gas cellular automaton (LGCA) as a model of cell migration together with a tensor characterization of heterogeneity of the microenvironment. LGCA have been originally introduced as discrete models of fluid dynamics; they allow modeling of migrating fluid particles in a straightforward manner. Tensor data of biological microenvironments may be provided by means of novel imaging techniques such as diffusion tensor imaging. The authors have provided estimates of the cell dispersion speed within a given environment.

4.6.2   Single-Cell-Based Models

In mathematical models, tumor cells can be either treated as discrete objects or approximated using a continuous density distribution (so-called mean field). Both approaches have their disadvantages and advantages, depending upon the phenomenon under consideration, and time- and space-scale at which the phenomenon is studied. Discrete models are capable of including detailed behavior of the cell, which includes cell–cell and cell–ECM interactions, cell proliferation, and biased and random migration. Continuum approaches are usually deterministic and do not cover the systems where stochastic fluctuations are important, or where the information is reflected only in stochastic fluctuations. In addition to the cellular automata models, among discrete models, single-cell-based models are distinguished. Typically, in these models location of the cell is not connected to a rigid rectangular grid point but is assigned arbitrarily—in this case we deal with the so-called off-lattice models. CA models can show good correspondence to real systems on mesoscopic length scale which is much larger than the cell diameter. Unfortunately, it is problematic to incorporate cell size changes, mechanical deformations or compressions of the cells in the CA models.

Drasdo and Hohme [42] have published an off-lattice single-cell-based tumor growth model. They study the spatio-temporal growth dynamics of two-dimensional tumor monolayers and three-dimensional tumor spheroids as a complementary tool to in vitro experiments. Each cell is represented by an individual object and parametrized by biophysical or kinetic parameters accessible experimentally. Model approaches each cell as an elastic, sticky particle of limited compressibility and deformability. Cells are capable of active migration, growth, and division. Cells that are in contact form adhesive bonds. Decreasing distance between cell centers increases the number of adhesive bonds, resulting in an increasing attractive interaction. If cell is in isolation, an increasing contact area is accompanied by an increasing deformation which results in a repulsive interaction. The model presumes either very fast or very slow lysis of dead cells by removing or not removing dead cells. The authors study which mechanisms at the microscopic level of individual cells may affect the macroscopic properties of a growing tumor. For example, the cell cycle time of proliferating cells can be extended as a consequence of mechanical pressure on the cell which, for sufficiently large cycle times, may even become apoptotic. They found growth kinetics and patterns at early growth stages to be remarkably robust. Quantitative comparisons between computer simulations and published experimental observations of monolayer cultures indicate that large tumors obey linear growth. Interestingly, depletion of the nutrient seems to determine mainly the size of the necrotic core but not the size of the tumor.

In contrast to Drasdo and Hohme, Rejniak [121] developed a model of tumor growth of a single fully deformable cell by using the immersed boundary method. This biomechanical approach couples a continuous description of a viscous incompressible cytoplasm with the dynamics of separate elastic cells, containing their own point nuclei, elastic plasma membranes with membrane receptors, and individually regulated cell processes. The boundary of each cell is modelled as a collection of massless springs applying a force to the fluid in which they are immersed. Both cell cytoplasm and the fluid outside the cells are modelled as the same homogeneous medium, but the lack of cell cytoskeleton and extracellular fibers is compensated by choosing a suitably higher fluid viscosity. Motion of the incompressible fluid is governed by the Navier–Stokes equations. All cells can sense mechanical and chemical signals from their immediate neighbourhood and can also send mechanical signals to the neighbouring cells using adherent links or cell membrane receptors. Cell-to-cell adhesion is represented by linear springs attached at the boundary points of two distinct cells if they are within their adhesion microenvironment of specific radius. The process of cell-to-cell attachment and detachment in real cells is very dynamic and depends on the distance between neighbouring cells and on the kind of receptors expressed on cell membranes. Cell proliferation is modelled by including all its mechanical phases, such as cell growth and shape elongation, formation of the contractile ring, the contractile furrow, and separation of daughter cells into two new entities with their own nuclei surrounded by the cytoplasm and enclosed by the plasma membrane. The model enables formation of clusters or sheets of cells that act together as one complex tissue. Developed model can be adapted to simulate a randomly growing cluster of tumor cells, nutrient-dependent tumor growth, or development of an intraductal tumor (see Fig. 4.4). The main advantage of the method over other existing cell-based models is the explicitly modelled cell membrane that allows defining all cell interactions at the level of cell membrane receptors. The model framework is flexible and allows many changes, which might increase the computational cost which is already high.

Another cell-based modeling approach is presented by Merks et al. [98]. He uses Cellular Potts Models to describe endothelial cell sprouting during angiogenesis. Authors focus on the behaviour of the cell which drives branch splitting during sprouting. The endothelial cell is modeled by a lattice-based Monte Carlo method. The method describes biological cells as spatially extended patches of identical lattice indices on a square or triangular lattice, where each index uniquely identifies, or labels, a single biological cell. This method enables to model the explicit shape of the cell, including pseudopod protrusions. The advantage of this method over other cell-based methods is that cell is not represented as a point particle or fixed-size sphere and allows distinguishing specific regions of cell membrane. In simulations the authors found branching instabilities, which result from contact inhibition of chemotaxis. They suggest that changes in cell adhesivity are not significant provided that contact-inhibition persists.

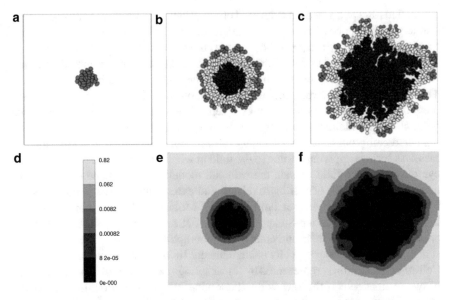

Fig. 4.4 Development of nutrient-driven finger-like tumor: (**a**)–(**c**) spatial distribution of sub-populations of tumor cells: necrotic—*black*, proliferating—*grey*, viable—*white*; (**e**)–(**f**) spatial distribution of nutrients corresponding to cases (**b**)–(**c**). Figure reprinted with the courtesy of Elsevier Science Ltd. [121]

4.6.3 Models Containing Therapeutic Actions

Recent studies by Bravo and Axelrod [26] show cell dynamics of normal human colon crypts by means of the agent-based model. They use stochastic model of the form of an on-lattice CA model. What is important, the model has been calibrated using measurements of human biopsy specimens. Model reproduced the behavior of biological crypts, including monoclonal conversion by neutral drift, formation of adenomas by initiating mutations at various locations along the crypt axis, and robust recovery after exposure to a cytotoxic agent. Simulations of different durations of chemotherapy showed the collateral damage to healthy cells.

Ribba et al. [123] have developed the hybrid cellular automata (HCA) modeling framework, a useful approach to deal with biological problems, particularly these involving multiple spatial and temporal scales. Authors showed application of HCA for assessing chemotherapy treatment for non-Hodgkin's lymphoma. Except the CA model for the cells, they model the vascular network as a honeycomb-like structure, and incorporate PK/PD features applicable to doxorubicin and cyclophosphamide drugs. They found that blood flow heterogeneity is the key factor in treatment simulations. They conclude that in order for treatment to be efficient, additional drug cycles must be administered before the tumor can enter the unstable stage of regrowth.

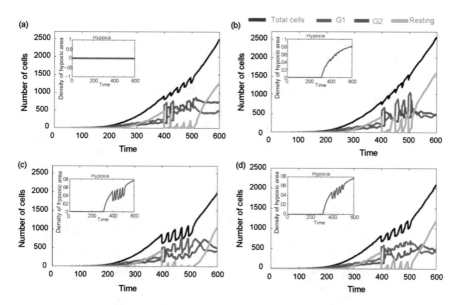

Fig. 4.5 Number of cells under various conditions that influence the radiation damage after the irradiation. Figure reprinted with the courtesy of Plos [115]. (**a**) Number of cells in well oxygenated environment. (**b**) Number of cells with no cell-cycle delay after radiation. (**c**) Number of cells with no phase specific sensitivity. (**d**) Number of cells with no dna repair

Another hybrid mathematical model was presented by Powathil et al. [115] in 2013. Model incorporates both individual cell behaviour through the cell-cycle and effects of the changing microenvironment through oxygen dynamics to study the multiple effects of radiation therapy. To study the spatio-temporal growth of cancer cells and their response to radiotherapy and chemotherapy they use of CA model (see Fig. 4.5). Cell-cycle dynamics within each cell was modeled by ODE equations from Tyson and Novak [153]. Model also contains effects of chemotherapy and radiotherapy, including the cell-cycle phase-specific radiosensitivity. Chemotherapeutic drugs act on rapidly proliferating cells. The distribution of chemotherapeutic drug is modelled by a reaction-diffusion equation. The survival probability of cells after they are irradiated is calculated by means of the LQ model where oxygenation status of the tissue is considered to have impact of the therapy effectiveness. Authors have studied various combinations of multiple doses of cell-cycle dependent chemotherapies and radiation therapy. They showed that scheduling the chemotherapeutic drugs can significantly affect the cell-cycle and oxygen heterogeneities of the tumor mass.

Lopez Alfonso et al. [88] focused on the radiotherapy treatment. They used a three-dimensional CA model together with the LQ model for estimation of the surviving fraction of cells after a radiation. They have studied different radiation dose distributions in heterogeneous tumors composed with ordinary cancer cells and cancer stem cells. The inner tumor regions where more radioresistant tumor

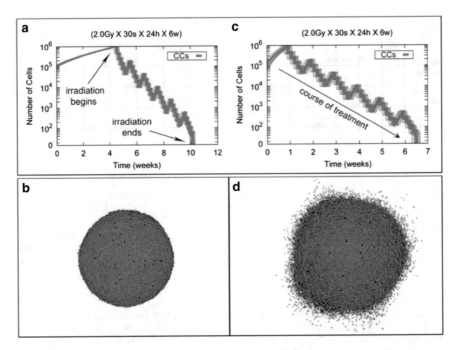

Fig. 4.6 Standard radiotherapy treatment in a homogeneous tumor for the low and high migration cases. Cell growth curves are shown corresponding to homogeneous tumor growth for the low (A) and high (C) migration cases. A homogeneous treatment corresponding to 30 sessions of 2.0 Gy each is delivered. In (B) and (D) tumor stages are represented when radiation therapy is started (about 10^6 cells in total) for the low and high migration cases respectively. Depicted in dark and light green are proliferating and quiescent cells. Dead cells are represented in black. Figure reprinted with the courtesy of Plos [88]

cell phenotype remains confined are shown to strongly depend on cancer stem cell cycle duration and their probability of asymmetric division. Boosting radiation in the region occupied by the more radioresistant tumor cell phenotype can enhance tumor control. Simulations of standard radiotherapy treatment in a homogeneous tumor for the low and high migration cases are presented in Fig. 4.6.

References

1. J.L. Abkowitz, S.N. Catlin, P. Guttorp, Evidence that hematopoiesis may be a stochastic process in vivo. Nat. Med. **2**(2), 190–197 (1996)
2. T. Alarcon, H.M. Byrne, P.K. Maini, A cellular automaton model for tumour growth in inhomogeneous environment. J. Theor. Biol. **225**(2), 257–274 (2003)
3. D. Ambrosi, L. Preziosi, On the closure of mass balance models for tumor growth. Math. Models Methods Appl. Sci. **12**, 737–754 (2002)

4. A.R. Anderson, M.J. Chaplain, Continuous and discrete mathematical models of tumor-induced angiogenesis. Bull. Math. Biol. **60**, 857–899 (1998)
5. R.P. Araujo, D.L.S. McElwain, A history of the study of solid tumour growth: the contribution of mathematical modelling. Bull. Math. Biol. **66**, 1039–1091 (2004)
6. R.P. Araujo, D.L.S. McElwain, A mixture theory for the genesis of residual stresses in growing tissues i: a general formulation. SIAM J. Appl. Math. **65**, 1261–1284 (2005)
7. R.P. Araujo, D.L.S. McElwain, A mixture theory for the genesis of residual stresses in growing tissues ii: solutions to the biphasic equations for a multicell spheroid. SIAM J. Appl. Math. **66**, 447–467 (2005)
8. O. Arino, M. Kimmel, Stability analysis of models of cell production systems. Math. Model. **7**(9–12), 1269–1300 (1986)
9. O. Arino, M. Kimmel, Asymptotic analysis of a cell cycle model based on unequal division. SIAM J. Appl. Math. **47**(1), 128–145 (1987)
10. O. Arino, M. Kimmel, Comparison of approaches to modeling of cell population dynamics. SIAM J. Appl. Math. **53**(5), 1480–1504 (1993)
11. N. Awaya, H. Uchida, Y. Miyakawa, K. Kinjo, H. Matsushita, H. Nakajima, Y. Ikeda, M. Kizaki, Novel variant isoform of G-CSF receptor involved in induction of proliferation of FDCP-2 cells: relevance to the pathogenesis of myelodysplastic syndrome. J. Cell. Physiol. **191**(3), 327–335 (2002)
12. A. Barranco-Mendoza, C. Clem, A. Gupta, P. Fizzano, M. Guillaud, Predicting the development of pre-invasive lesions from biopsies. Arch. Control Sci. **9**, 25–40 (1999)
13. V. Becker, M. Schilling, J. Bachmann, U. Baumann, A. Raue, T. Maiwald, J. Timmer, U. Klingmuller, Covering a broad dynamic range: information processing at the erythropoietin receptor. Sci. Signal. **328**(5984), 1404–1408 (2010)
14. R. Beekman, I.P. Touw, G-CSF and its receptor in myeloid malignancy. Blood **115**(25), 5131–5136 (2010)
15. R. Beekman, M.G. Valkhof, M.A. Sanders, P.M. van Strien, J.R. Haanstra, L. Broeders, W.M. Geertsma-Kleinekoort, A.J. Veerman, P.J. Valk, R.G. Verhaak, B. Lowenberg, I.P. Touw, Sequential gain of mutations in severe congenital neutropenia progressing to acute myeloid leukemia. Blood **119**(22), 5071–5077 (2012)
16. N. Bellomo, M. Delitala, From the mathematical kinetic, and stochastic game theory to modelling mutations, onset, progression and immune competition of cancer cells. Phys. Life Rev. **5**, 183–206 (2008)
17. S. Bernard, J. Belair, M.C. Mackey, Oscillations in cyclical neutropenia: new evidence based on mathematical modeling. J. Theor. Biol. **223**(3), 283–298 (2003)
18. R. Bertolusso, M. Kimmel, Modeling spatial effects in early carcinogenesis: stochastic versus deterministic reaction-diffusion systems. Math. Model. Nat. Phenom. **7**(01), 245–260 (2012)
19. A. Bertuzzi, A. Gandolfi, Recent views on the cell cycle structure. Bull. Math. Biol. **45**, 605–616 (1983)
20. A. Bertuzzi, A. Fasano, A. Gandolfi, D. Marangi, Cell kinetics in tumour cords studied by a model with variable cell cycle length. Math. Biosci. **177/178**, 103–125 (2002)
21. A. Bertuzzi, A. Fasano, A. Gandolfi, A mathematical model for tumor cords incorporating the flow of interstitial fluid. Math. Models Methods Appl. Sci. **15**, 1735–1777 (2005)
22. A. Bertuzzi, A. Fasano, A. Gandolfi, C. Sinisgalli, Cell resensitization after delivery of a cycle-specific anticancer drug and effect of dose splitting: learning from tumour cords. J. Theor. Biol. **244**, 388–399 (2007)
23. F. Billy, J. Clairambault, Designing proliferating cell population models with functional targets for control by anti-cancer drugs. Discret. Cont. Dyn. Syst. **18**(4), 865–889 (2013)
24. F. Billy, J. Clairambault, O. Fercoq, S. Gaubertt, T. Lepoutre, T. Ouillon, S. Saito, Synchronisation and control of proliferation in cycling cell population models with age structure. Math. Comput. Simul. **96**, 66–94 (2014)
25. I. Bozic, T. Antal, H. Ohtsuki, H. Carter, D. Kim, S. Chen, R. Karchin, K.W. Kinzler, B. Vogelstein, M.A. Nowak, Accumulation of driver and passenger mutations during tumor progression. Proc. Natl. Acad. Sci. **107**(43), 18545–18550 (2010)

26. R. Bravo, D.A. Axelrod, A calibrated agent-based computer model of stochastic cell dynamics in normal human colon crypts useful for in silico experiments. Theor. Biol. Med. Model. **10**(1), 66 (2013)
27. D.J. Brenner, The linear-quadratic model is an appropriate methodology for determining isoeffective doses at large doses per fraction. Semin. Radiat. Oncol. **18**(4), 234–239 (2008)
28. D. Bresch, T. Colin, E. Grenier, B. Ribba, O. Saut, Computational modeling of solid tumor growth: the avascular stage. SIAM J. Sci. Comput. **32**, 2321–2344 (2010)
29. A. Brock, H. Chang, S. Huang, Non-genetic heterogeneity—a mutation-independent driving force for the somatic evolution of tumours. Nat. Rev. Genet. **10**(5), 336–342 (2009)
30. H. Byrne, L. Preziosi, Modelling solid tumour growth using the theory of mixtures. Math. Med. Biol. **20**, 341–366 (2003)
31. H.M. Byrne, J.R. King, D.L.S. McElwain, L. Preziosi, A two-phase model of solid tumour growth. Appl. Math. Lett. **16**, 567–573 (2003)
32. H.C. Chang, M. Hemberg, M. Barahona, D.E. Ingber, S. Huang, Transcriptome-wide noise controls lineage choice in mammalian progenitor cells. Nature **453**(7194), 544–547 (2008)
33. M.A. Chaplain, Avascular growth, angiogenesis and vascular growth in solid tumours: the mathematical modelling of the stages of tumour development. Math. Comput. Model. **23**, 47–87 (1996)
34. M.A. Chaplain, A.R. Anderson, Mathematical modelling, simulation and prediction of tumour-induced angiogenesis. Invasion Metastasis **16**, 222–234 (1996)
35. M.A.J. Chaplain, L. Graziano, L. Preziosi, Mathematical modelling of the loss of tissue compression responsiveness and its role in solid tumour development. Math. Med. Biol. **23**, 197–229 (2006)
36. A.A. Cohen, N. Geva-Zatorsky, E. Eden, M. Frenkel-Morgenstern, I. Issaeva, A. Sigal, R. Milo, C. Cohen-Saidon, Y. Liron, Z. Kam, L. Cohen, T. Danon, N. Perzov, U. Alon, Dynamic proteomics of individual cancer cells in response to a drug. Science **322**(5907), 1511–1516 (2008)
37. A. Dawson, T. Hillen, Derivation of the tumour control probability (TCP) from a cell cycle model. Comput. Math. Methods Med. **7**(2–3), 121–141 (2006)
38. O. Diekmann, Modeling and analysing physiologically structured populations, in *Mathematics Inspired by Biology*. Lecture Notes in Mathematics, vol. 1714 (Springer, New York, 1999), pp. 1–37
39. F. Dong, M. van Paassen, C. van Buitenen, L.H. Hoefsloot, B. Lowenberg, I.P. Touw, A point mutation in the granulocyte colony-stimulating factor receptor (G-CSF-R) gene in a case of acute myeloid leukemia results in the overexpression of a novel G-CSF-R isoform. Blood **85**(4), 902–911 (1995)
40. S. Dormann, A. Deutsch, Modeling of self-organized avascular tumor growth with a hybrid cellular automaton. In Silico Biol. **2**, 393–406 (2002)
41. M. Doumic, A. Marciniak-Czochra, B. Perthame, J.P. Zubelli, A structured population model of cell differentiation. SIAM J. Appl. Math. **71**(6), 1918–1940 (2011)
42. D. Drasdo, S. Hohme, A single-cell-based model of tumor growth in vitro: monolayers and spheroids. Phys. Biol. **2**, 133–147 (2005)
43. J. Dyson, R. Villella-Bressan, G.F. Webb, A maturity structured model of a population of proliferating and quiescent cells. Arch. Control Sci. **9**, 201–225 (1999)
44. J. Dyson, R. Villella-Bressan, G.F. Webb, A nonlinear age and maturity structured model of population dynamics: I. Basic theory. J. Math. Anal. Appl. **242**(1), 93–104 (2000)
45. J. Dyson, R. Villella-Bressan, G.F. Webb, Asynchronous exponential growth in an age structured population of proliferating and quiescent cells. Math. Biosci. **177–178**, 73–83 (2002)
46. M.B. Elowitz, A.J. Levine, E.D. Siggia, P.S. Swain, Stochastic gene expression in a single cell. Sci. Signal. **297**(5584), 1183–1186 (2002)
47. H. Enderling, M.A. Chaplain, A.R.A. Anderson, J.S. Vaidya, A mathematical model of breast cancer development, local treatment and recurrence. J. Theor. Biol. **246**(2), 245–259 (2007)

48. G.B. Ermentrout, L. Edelstein-Keshet, Cellular automata approaches to biological modeling. J. Theor. Biol. **160**(1), 97–133 (1993)
49. C. Foley, M.C. Mackey, Dynamic hematological disease: a review. J. Math. Biol. **58**(1–2), 285–322 (2009)
50. S.J. Franks, H.M. Byrne, J.R. King, J.C.E. Underwood, C.E. Lewis, Modelling the early growth of ductal carcinoma in situ of the breast. J. Math. Biol. **47**, 424–452 (2003)
51. T.S. Gardner, C.R. Cantor, J.J. Collins, Construction of a genetic toggle switch in *Escherichia coli*. Nature **403**(6767), 339–342 (2000)
52. R.A. Gatenby, E.T. Gawlinski, A.F. Gmitro, B. Kaylor, R.J. Gillies, Acid-mediated tumor invasion: a multidisciplinary study. Cancer Res. **66**, 5216–5223 (2006)
53. M. Germeshausen, J. Skokowa, M. Ballmaier, C. Zeidler, K. Welte, G-CSF receptor mutations in patients with congenital neutropenia. Curr. Opin. Hematol. **15**(4), 332–337 (2008)
54. P. Getto, A. Marciniak-Czochra, Y. Nakata, Global dynamics of two-compartment models for cell production systems with regulatory mechanisms. Math. Biosci. **245**(2), 258–268 (2013)
55. T. Glaubach, S.J. Corey, From famine to feast: sending out the clones. Blood **119**(22), 5063–5064 (2012)
56. H.P. Greenspan, Models for the growth of a solid tumor by diffusion. Stud. Appl. Math. **4**, 317–340 (1972)
57. P.B. Gupta, C.M. Fillmore, G. Jiang, S.D. Shapira, K. Tao, C. Kuperwasser, E.S. Lander, Stochastic state transitions give rise to phenotypic equilibrium in populations of cancer cells. Cell **146**(4), 633–644 (2011)
58. P. Gwiazda, G. Jamroz, A. Marciniak-Czochra, Models of discrete and continuous cell differentiation in the framework of transport equation. SIAM J. Math. Anal. **44**(2), 1103–1133 (2012)
59. P. Haccou, P. Jagers, V.A. Vatutin, *Branching Processes: Variation, Growth, and Extinction of Populations*, vol. 5 (Cambridge University Press, Cambridge, 2005)
60. D. Hanahan, R.A. Weinberg, The hallmarks of cancer. Cell **100**, 57–70 (2000)
61. D. Hanahan, R.A. Weinberg, The hallmarks of cancer: the next generation. Cell **144**, 646–674 (2011)
62. L.E. Harnevo, Z. Agur, Drug resistance as a dynamic process in a model for multistep gene amplification under various levels of selection stringency. Cancer Chemother. Pharmacol. **30**(6), 469–476 (1992)
63. H. Hatzikirou, A. Deutsch, Cellular automata as microscopic models of cell migration in heterogeneous environments. Curr. Top. Dev. Biol. **81**, 401–434 (2008)
64. D.J. Hicklin, L.M. Ellis, Role of the vascular endothelial growth factor pathway in tumor growth and angiogenesis. J. Clin. Oncol. **23**(5), 1011–1027 (2005)
65. T. Hillen, K.J. Painter, A user's guide to PDE models for chemotaxis. J. Math. Biol. **58**, 183–217 (2009)
66. P. Hinow, P. Gerlee, L.J. McCawley, V. Quaranta, M. Ciobanu, S. Wang, J.M. Graham, B.P. Ayati, J. Claridge, K.R. Swanson, M. Loveless, A.R.A. Anderson, A spatial model of tumor-host interaction: application of chemotherapy. Math. Biosci. Eng. **6**(3), 521–546 (2009)
67. S. Huang, Systems biology of stem cells: three useful perspectives to help overcome the paradigm of linear pathways. Philos. Trans. R. Soc. Lond. Ser. B Biol. Sci. **366**(1575), 2247–2259 (2011)
68. J. Jaruszewicz, P.J. Zuk, T. Lipniacki, Type of noise defines global attractors in bistable molecular regulatory systems. J. Theor. Biol. **317**, 150–151 (2013)
69. Y. Kam, K.A. Rejniak, A.R. Anderson, Cellular modeling of cancer invasion: integration of in silico and in vitro approaches. J. Cell. Physiol. **227**, 431–438 (2012)
70. A.R. Kansal, S. Torquato, G.R. Harsh IV, E.A. Chiocca, T.S. Deisboeck, Simulated brain tumor growth dynamics using a three-dimensional cellular automaton. J. Theor. Biol. **203**(4), 367–382 (2000)
71. P.S. Kim, P.P. Lee, D. Levy, A PDE model for imatinib-treated chronic myelogenous leukemia. Bull. Math. Biol. **70**(7), 1994–2016 (2008)

72. M. Kim, R.J. Gillies, K.A. Rejniak, Current advances in mathematical modeling of anti-cancer drug penetration into tumor tissues. Front. Oncol. **3**, 278 (2013)
73. M. Kimmel, D.E. Axelrod, Mathematical models of gene amplification with applications to cellular drug resistance and tumorigenicity. Genetics **125**(3), 633–644 (1990)
74. M. Kimmel, D.E. Axelrod, *Branching Processes in Biology*, extended, 2nd edn. (Springer, New York, 2015)
75. M. Kimmel, S. Corey, Stochastic hypothesis of transition from inborn neutropenia to AML: interactions of cell population dynamics and population genetics. Front. Oncol. **3**, 89 (2013)
76. M. Kimmel, Z. Darzynkiewicz, O. Arino, F. Traganos, Analysis of a cell cycle model based on unequal division of metabolic constituents to daughter cells during cytokinesis. J. Theor. Biol. **110**(4), 637–664 (1984)
77. N. Kronik, Y. Kogan, M. Elishmereni, K. Halevi-Tobias, S. Vuk-Pavlovic, Z. Agur, Predicting outcomes of prostate cancer immunotherapy by personalized mathematical models. PLoS ONE **5**(12), e15482 (2010)
78. A.D. Lander, K.K. Gokoffski, F.Y.M. Wan, Q. Nie, A.L. Calof, Cell lineages and the logic of proliferative control. PLoS Biol. **7**(1), e1000015 (2009)
79. P. Laslo, C.J. Spooner, A. Warmflash, D.W. Lancki, H.J. Lee, R. Sciammas, B.N. Gantner, A.R. Dinner, H. Singh, Multilineage transcriptional priming and determination of alternate hematopoietic cell fates. Cell **126**(4), 755–766 (2006)
80. P. Laslo, J.M.R. Pongubala, D.W. Lancki, H. Singh, Gene regulatory networks directing myeloid and lymphoid cell fates within the immune system, in *Seminars in Immunology*, vol. 20, no. 4 (Academic, London, 2008), pp. 228–235
81. M. Laurent, J. Deschatrette, C.M. Wolfrom, Unmasking chaotic attributes in time series of living cell populations. PLoS ONE **5**(2), e9346 (2010)
82. H.A. Levine, B.P. Sleeman, N. Nilsen-Hamilton, A mathematical modeling for the roles of pericytes and macrophages in the initiation of angiogenesis I. The role of protease inhibitors in preventing angiogenesis. Math. Biosci. **168**, 75–115 (2000)
83. T. Ley et al., The cancer genome atlas research network. Genomic and epigenomic landscapes of adult de novo acute myeloid leukemia. N. Engl. J. Med. **368**(22), 2059–2074 (2013)
84. T. Lipniacki, P. Paszek, A. Marciniak-Czochra, A.R. Brasier, M. Kimmel, Transcriptional stochasticity in gene expression. J. Theor. Biol. **238**(2), 348–367 (2006)
85. W.C. Lo, C.S. Chou, K.K. Gokoffski, F.Y.M. Wan, A.D. Lander, A.L. Calof, Q. Nie, Feedback regulation in multistage cell lineages. Math. Biosci. Eng. **6**(1), 59–82 (2009)
86. A. Loinger, A. Lipshtat, N.Q. Balaban, O. Biham, Stochastic simulations of genetic switch systems. Phys. Rev. E **75**(2), 021904 (2009)
87. M. Loose, G. Swiers, R. Patient, Transcriptional networks regulating hematopoietic cell fate decisions. Curr. Opin. Hematol. **14**(4), 307–314 (2007)
88. J.C. Lopez Alfonso, N. Jagiella, L. Nunez, M.A. Herrero, D. Drasdo, Estimating dose painting effects in radiotherapy: a mathematical model. PLoS ONE **9**(2), e89380 (2014)
89. A. Lorz, T. Lorenzi, J. Clairambault, A. Escargueil, B. Perthame, Modeling the effects of space structure and combination therapies on phenotypic heterogeneity and drug resistance in solid tumors. Bull. Math. Biol. **77**, 1–22 (2015)
90. J.S. Lowengrub, H.B. Frieboes, F. Jin, Y.-L. Chuang, X. Li, P. Macklin, S.M. Wise, V. Cristini, Nonlinear modelling of cancer: bridging the gap between cells and tumours. Nonlinearity **23**, R1–R9 (2010)
91. P. Macklin, J. Lowengrub, Nonlinear simulation of the effect of microenvironment on tumor growth. J. Theor. Biol. **245**, 677–704 (2007)
92. P. Macklin, S. McDougall, A.R. Anderson, M.A. Chaplain, V. Cristini, J. Lowengrub, Multiscale modelling and nonlinear simulation of vascular tumour growth. J. Math. Biol. **58**, 765–798 (2009)
93. E. Mamontov, A. Koptioug, K. Psiuk-Maksymowicz, The minimal, phase-transition model for the cell-number maintenance by the hyperplasia-extended homeorhesis. Acta Biotheor. **54**, 61–101 (2006)

94. A. Marciniak-Czochra, M. Kimmel, Reaction-diffusion model of early carcinogenesis: the effects of influx of mutated cells. Math. Model. Nat. Phenom. **3**(7), 90–114 (2008)
95. A. Marciniak-Czochra, T. Stiehl, W. Wagner, Modeling of replicative senescence in hematopoietic development. Aging (Albany NY) **1**(8), 723–732 (2009)
96. A. Marciniak-Czochra, T. Stiehl, A.D. Ho, W. Jager, W. Wagner, Modeling of asymmetric cell division in hematopoietic stem cells-regulation of self-renewal is essential for efficient repopulation. Stem Cells Dev. **18**(3), 377–386 (2009)
97. A. McKendrick, Applications of mathematics to medical problems. Proc. Edinb. Math. Soc. **44**, 98–130 (1926)
98. R.M.H. Merks, E.D. Perryn, A. Shirinifard, J.A. Glazier, Contact-inhibited chemotaxis in de novo and sprouting blood-vessel growth. PLoS Comput. Biol. **4**(9), e1000163 (2008)
99. J.L. Michaels, V. Naudot, L.S. Liebovitch, Dynamic stabilization in the PU1-GATA1 circuit using a model with time-dependent kinetic change. Bull. Math. Biol. **73**(9), 2132–2151 (2011)
100. F. Milde, M. Bergdorf, P. Koumoutsakos, A hybrid model for three-dimensional simulations of Sprouting Angiogenesis. Biophys. J. **95**(7), 3146–3160 (2008)
101. S.H. Moolgavkar, A.G. Knudson, Mutation and cancer: a model for human carcinogenesis. J. Nat. Cancer Inst. **66**(6), 1037–1052 (1981)
102. K.A. Moore, I.R. Lemischka, Stem cells and their niches. Science **311**(5769), 1880–1885 (2006)
103. S.J. Morrison, J. Kimble, Asymmetric and symmetric stem-cell divisions in development and cancer. Nature **441**(7097), 1068–1074 (2006)
104. D. Muzzey, A. van Oudenaarden, When it comes to decisions, myeloid progenitors crave positive feedback. Cell **126**(4), 650–652 (2006)
105. Y. Nakata, P. Getto, A. Marciniak-Czochra, T. Alarcon, Stability analysis of multi-compartment models for cell production systems. J. Biol. Dyn. **6**(Suppl. 1), 2–18 (2012)
106. P. Nurse, P. Thuriaux, Controls over the timing of DNA replication during the cell cycle of fission yeast. Exp. Cell Res. **107**, 365–375 (1977)
107. M. Ogawa, Stochastic model revisited. Int. J. Hematol. **69**(1), 2–5 (1999)
108. M.R. Owen, T. Alarcon, P.K. Maini, H.M. Byrne, Angiogenesis and vascular remodelling in normal and cancerous tissues. J. Math. Biol. **58**, 689–721 (2009)
109. K.J. Painter, T. Hillen, Mathematical modelling of glioma growth: the use of diffusion tensor imaging (DTI) data to predict the anisotropic pathways of cancer invasion. J. Theor. Biol. **323**, 25–39 (2013)
110. B. Parkin, P. Ouillette, Y. Li, J. Keller, C. Lam, D. Roulston, C. Li, K. Shedden, S.N. Malek, Clonal evolution and devolution after chemotherapy in adult acute myelogenous leukemia. Blood **121**(2), 369–377 (2013)
111. A.A. Patel, E.T. Gawlinski, S.K. Lemieux, R.A. Gatenby, A cellular automaton model of early tumor growth and invasion: the effects of native tissue vascularity and increased anaerobic tumor metabolism. J. Theor. Biol. **213**(3), 315–331 (2001)
112. D. Peixoto, D. Dingli, J.M. Pacheco, Modelling hematopoiesis in health and disease. Math. Comput. Model. **53**(7), 1546–1557 (2011)
113. L. Pelkmans, Using cell-to-cell variability—a new era in molecular biology. Science **336**(6080), 425–426 (2012)
114. C.P. Please, G.J. Pettet, D.L.S. McElwain, A new approach to modelling the formation of necrotic regions in tumours. Appl. Math. Lett. **11**, 89–94 (1998)
115. G.G. Powathil, D.J. Adamson, M.A. Chaplain, Towards predicting the response of a solid tumour to chemotherapy and radiotherapy treatments: clinical insights from a computational model. PLOS Comput. Biol. **9**(7), e1003120 (2013)
116. L. Preziosi, A. Tosin, Multiphase modelling of tumour growth and extracellular matrix interaction: mathematical tools and applications. J. Math. Biol. **58**, 625–656 (2009)
117. K. Psiuk-Maksymowicz, Multiphase modelling of desmoplastic tumour growth. J. Theor. Biol. **329**, 52–63 (2013)
118. K. Psiuk-Maksymowicz, E. Mamontov, Homeorhesis-based modelling and fast numerical analysis for oncogenic hyperplasia under radiotherapy. Math. Comput. Model. **47**, 580–596 (2008)

119. T. Radivoyevitch, L. Hlatky, J. Landaw, R.K. Sachs, Quantitative modeling of chronic myeloid leukemia: insights from radiobiology. Blood **119**(19), 4363–4371 (2012)
120. J.M. Raser, E.K. OŚhea, Noise in gene expression: origins, consequences, and control. Science **309**(5743), 2010–2013 (2005)
121. K.A. Rejniak, An immersed boundary framework for modelling the growth of individual cells: an application to the early tumour development. J. Theor. Biol. **247**(1), 186–204 (2007)
122. A. Raue, V. Becker, U. Klingmuller, J. Timmer, Identifiability and observability analysis for experimental design in nonlinear dynamical models. Chaos: Interdiscip. J. Nonlinear Sci. **20**(4), 045105–045105 (2010)
123. B. Ribba, T. Alarcon, K. Marron, P.K. Maini, Z. Agur, The use of hybrid cellular automaton models for improving cancer therapy, in *ACRI 2004*, ed. by P.M.A. Sloot, B. Chopard, A.G. Hoekstra. Lecture Notes in Computer Science, vol. 3305 (Springer, Berlin/Heidelberg, 2004), pp. 444–453
124. B. Ribba, T. Colin, S. Schnell, A multiscale mathematical model of cancer, and its use in analyzing irradiation therapies. Theor. Biol. Med. Model. **3**, 7 (2006)
125. B. Ribba, O. Saut, T. Colin, D. Bresch, E. Grenier, J.P. Boissel, A multiscale mathematical model of avascular tumor growth to investigate the therapeutic benefit of anti-invasive agents. J. Theor. Biol. **243**, 532–541 (2006)
126. R. Rockne, E.C. Alvord Jr, J.K. Rockhill, K.R. Swanson, A mathematical model for brain tumor response to radiation therapy. J. Math. Biol. **58**(4–5), 561–578 (2009)
127. R. Rockne, J.K. Rockhill, M. Mrugala, A.M. Spence, I. Kalet, K. Hendrickson, A. Lai, T. Cloughesy, E.C. Alvord Jr, K.R. Swanson, Predicting the efficacy of radiotherapy in individual glioblastoma patients in vivo: a mathematical modeling approach. Phys. Med. Biol. **55**(12), 3271–3285 (2010)
128. I. Roeder, M. Horn, I. Glauche, A. Hochhaus, M.C. Mueller, M. Loeffler, Dynamic modeling of imatinib-treated chronic myeloid leukemia: functional insights and clinical implications. Nat. Med. **12**(10), 1181–1184 (2006)
129. T. Roose, S.J. Chapman, P.K. Maini, Mathematical models of avascular tumor growth. SIAM Rev. **49**, 179–208 (2007)
130. P.S. Rosenberg, B.P. Alter, A.A. Bolyard, M.A. Bonilla, L.A. Boxer, B. Cham, C. Fier, M. Freedman, G. Kannourakis, S. Kinsey, B. Schwinzer, C. Zeidler, K. Welte, D.C. Dale, Severe chronic neutropenia international Registry. The incidence of leukemia and mortality from sepsis in patients with severe congenital neutropenia receiving long-term G-CSF therapy. Blood **107**(12), 4628–4635 (2006)
131. P.S. Rosenberg, C. Zeidler, A.A. Bolyard, B.P. Alter, M.A. Bonilla, L.A. Boxer, Y. Dror, S. Kinsey, D.C. Link, P.E. Newburger, A. Shimamura, K. Welte, D.C. Dale, Stable long-term risk of leukaemia in patients with severe congenital neutropenia maintained on G-CSF therapy. Br. J. Haematol. **150**(2), 196–199 (2010)
132. M.S. Samoilov, G. Price, A.P. Arkin, From fluctuations to phenotypes: the physiology of noise. Sci. Signal. **366**, re17 (2006)
133. T. Schroeder, Asymmetric cell division in normal and malignant hematopoietic precursor cells. Cell Stem Cell **1**(5), 479–481 (2007)
134. T. Schroeder, Long-term single-cell imaging of mammalian stem cells. Nat. Methods **8**(4s), S30–S35 (2011)
135. M. Sehl, H. Zhou, J.S. Sinsheimer, K.L. Lange, Extinction models for cancer stem cell therapy. Math. Biosci. **234**(2), 132–146 (2011)
136. J.A. Sethian, A fast marching level set method for monotonically advancing fronts. Proc. Natl. Acad. Sci. **93**(4), 1591–1595 (1996)
137. J.A. Sherratt, Traveling wave solutions of a mathematical model for tumor encapsulation. SIAM J. Appl. Math. **60**, 392–407 (1999)
138. J.A. Sherratt, Predictive mathematical modeling in metastasis. Methods Mol. Med. **57**, 309–315 (2001)
139. A. Sigal, R. Milo, A. Cohen, N. Geva-Zatorsky, Y. Klein, Y. Liron, N. Rosenfeld, T. Danon, N. Perzov, U. Alon, Variability and memory of protein levels in human cells. Nature **444**(7119), 643–646 (2006)

140. W.K. Sinclair, The shape of radiation survival curves of mammalian cells cultured in vitro, in *Biophysical Aspects of Radiation Quality*. Technical Reports Series, vol. 58 (International Atomic Energy Agency, Vienna, 1966), pp. 21–43

141. B. Snijder, L. Pelkmans, Origins of regulated cell-to-cell variability. Nat. Rev. Mol. Cell Biol. **12**(2), 119–125 (2011)

142. T. Stiehl, A. Marciniak-Czochra, Characterization of stem cells using mathematical models of multistage cell lineages. Math. Comput. Model. **53**(7), 1505–1517 (2011)

143. T. Stiehl, A. Marciniak-Czochra, Mathematical modeling of leukemogenesis and cancer stem cell dynamics. Math. Model. Nat. Phenom. **7**(1), 166–202 (2012)

144. T. Stiehl, N. Baran, A.D. Ho, A. Marciniak-Czochra, Clonal selection and therapy resistance in acute leukaemias: mathematical modelling explains different proliferation patterns at diagnosis and relapse. J. R. Soc. Interface **11**(94), 20140079 (2014)

145. T. Stiehl, A.D. Ho, A. Marciniak-Czochra, The impact of CD34+ cell dose on engraftment after SCTs: personalized estimates based on mathematical modeling. Bone Marrow Transplant. **49**(1), 30–37 (2014)

146. T. Stiehl, N. Baran, A.D. Ho, A. Marciniak-Czochra, Cell division patterns in acute myeloid leukemia stem-like cells determine clinical course: a model to predict patient survival. Cancer Res. **75**(6), 940–949 (2015)

147. K.R. Swanson, C. Bridge, J.D. Murray, E.C. Alvord Jr., Virtual and real brain tumors: using mathematical modeling to quantify glioma growth and invasion. J. Neurol. Sci. **216**, 1–10 (2003)

148. H. Takizawa, S. Boettcher, M.G. Manz, Demand-adapted regulation of early hematopoiesis in infection and inflammation. Blood **119**(13), 2991–3002 (2012)

149. K.S. Tieu, R.S. Tieu, J.A. Martinez-Agosto, M.E. Sehl, Stem cell niche dynamics: from homeostasis to carcinogenesis. Stem Cells Int. **2012**, 367567 (2012)

150. C. Tomasetti, B. Vogelstein, G. Parmigiani, Half or more of the somatic mutations in cancers of self-renewing tissues originate prior to tumor initiation. Proc. Natl. Acad. Sci. **110**(6), 1999–2004 (2013)

151. A. Traulsen, J.M. Pacheco, L. Luzzatto, D. Dingli, Somatic mutations and the hierarchy of hematopoiesis. Bioessays **32**(11), 1003–1008 (2010)

152. J.J. Tyson, K.B. Hannsgen, Cell growth and division: a deterministic/probabilistic model of the cell cycle. J. Math. Biol. **23**(2), 231–246 (1986)

153. J.J. Tyson, B. Novak, Regulation of the eukaryotic cell cycle: molecular antagonism, hysteresis, and irreversible transitions. J. Theor. Biol. **210**, 249–263 (2001)

154. J.M. Vose, J.O. Armitage, Clinical applications of hematopoietic growth factors. J. Clin. Oncol. **13**(4), 1023–1035 (1995)

155. T. Walenda, T. Stiehl, H. Braun, J. Frobel, A.D. Ho, T. Schroeder, T.W. Goecke, B. Rath, U. Germing, A. Marciniak-Czochra, W. Wagner, Feedback signals in myelodysplastic syndromes: increased self-renewal of the malignant clone suppresses normal hematopoiesis. PLoS Comput. Biol. **10**(4), e1003599 (2014)

156. J.P. Ward, J.R. King, Mathematical modelling of avascular tumour growth. IMA J. Math. Appl. Med. Biol. **14**, 36–69 (1997)

157. M. Wazewska-Czyzewska, A. Lasota. Mathematical models of the red cell system. Matematyta Stosowana **6**, 25–40 (1976)

158. G.F. Webb, Random transitions, size control, and inheritance in cell population dynamics. Math. Biosci. **85**(1), 71–91 (1987)

159. M. Welter, K. Bartha, H. Rieger, Vascular remodelling of an arterio-venous blood vessel network during solid tumour growth. J. Theor. Biol. **259**, 405–422 (2009)

160. Z.L. Whichard, C.A. Sarkar, M. Kimmel, S.J. Corey, Hematopoiesis and its disorders: a systems biology approach. Blood **115**(12), 2339–2347 (2010)

161. A. Wilson, E. Laurenti, G. Oser, R.C. van der Wath, W. Blanco-Bose, M. Jaworski, S. Offner, C.F. Dunant, L. Eshkind, E. Bockamp, P. Lio, H.R. Macdonald, A. Trumpp, Hematopoietic stem cells reversibly switch from dormancy to self-renewal during homeostasis and repair. Cell **135**(6), 1118–1129 (2008)

162. S. Wise, J. Kim, J.S. Lowengrub, Solving the regularized, strongly anisotropic Chan-Hilliard equation by an adaptive nonlinear multigrid method. J. Comput. Phys. **226**, 414–446 (2007)
163. S.M. Wise, J.S. Lowengrub, H.B. Frieboes, V. Cristini, Three-dimensional multispecies nonlinear tumor growth–I Model and numerical method. J. Theor. Biol. **253**, 524–543 (2008)
164. M. Wu, H.Y. Kwon, F. Rattis, J. Blum, C. Zhao, R. Ashkenazi, T.L. Jackson, N. Gaiano, T. Oliver, T. Reya, Imaging hematopoietic precursor division in real time. Cell Stem Cell **1**(5), 541–554 (2007)

Chapter 5
Signaling Pathways Dynamics and Cancer Treatment

Abstract The rapid development of the biological research techniques in recent years provides more and more high quality data. Current technology allows to observe not only the whole body, tissue, or cell but also what happens inside the single cell, for example the time change (dynamics) of the number and location of biomolecules such as proteins, and their properties such as phosphorylation. With this knowledge it becomes obvious that the intracellular interactions between various molecules are not straightforward but complex with many mutual dependencies and feedback loops. It becomes clear that for a better understanding of the networks and their dynamics a complex approach is required. This approach can be based on the methodology of system engineering. It involves construction of mathematical models of observed phenomena and then their analysis. Mathematical models may be deterministic, based on ordinary differential equations (ODEs) or stochastic based on reaction propensities. They cover the network of interactions of intracellular species called the signaling pathways. In this chapter we define signaling pathways, describe main reaction types and corresponding equations, describe the main numerical methods for simulation of deterministic and stochastic models, and discuss the basics of the signaling pathways model analysis including stability, sensitivity, and bifurcation analysis. At the end we present an example of the p53 signaling pathway model and discuss possible anticancer therapies using available control signals.

5.1 Deterministic and Stochastic Models of Regulatory Modules in Signaling Pathways

5.1.1 Signaling Pathways

A signaling pathway may be defined as a set of substrates such as proteins, transcripts, enzymes, genes, and receptors interacting with each other to control one or more of the cellular functions. A sequence of proteins activation and/or inactivation (especially protein kinases) could be considered as a means of information processing and development of an appropriate response to stimuli.

© Springer International Publishing Switzerland 2016
A. Świerniak et al., *System Engineering Approach to Planning Anticancer Therapies*, DOI 10.1007/978-3-319-28095-0_5

Signaling pathways usually have chain structure with positive and negative feedback and feedforward loops (regulatory modules). Signal transduction involves phosphorylations, acetylations, and ubiquitinations of proteins, for example phosphorylation of the IKK kinases by IKKK kinases in the NF-κB signaling pathway or Akt protein by PIP3 in the p53 signaling pathway.

Signaling pathways are composed of the following functional modules:

- Receptor activation and deactivation — their role is to receive the extracellular signal and transfer it into the cell. There are many signaling pathways which start from receptor activation, for example NF-κB [36]. Usually the probability of receptors activation is proportional to the activating cytokine concentration in the environment. Receptor deactivation may be spontaneous because of the cytokine detachment or endocytosis (internalization of the receptor into the cell), or it may depend on external factors.

- Gene activation and deactivation — gene activation occurs when a transcription factor attaches to the promoter region, which initiates gene transcription. Gene deactivation is caused by spontaneous or forced detachment of the transcription factor from the promoter region or attachment of the gene's repressor.

- Transcription and translation — an active gene produces mRNA of the corresponding protein. This ends when the gene is deactivated. As a result one active gene could produce several to several thousand mRNA copies. mRNA is then transported to the cytoplasm where it is translated in the ribosomes into the protein. Once again, one mRNA molecule may be used to produce from several to hundreds of protein molecules. The mechanism of transcription and translation includes powerful amplification; a single gene activation event may result in hundreds of thousands of protein molecules.

- Signaling cascades—a single particle of active kinase may be a catalyst for hundreds of downstream kinases activation. If these are catalysts for further activations and so forth, the few active kinases at the beginning of the chain will result in hundreds of thousands of active kinases at the end of the chain. The chain of activations leading to strong amplification of the signal is called the signaling cascade.

- Negative and positive feedback loops—signal transduction in cells is not unidirectional. Usually in the regulatory modules such as the p53 regulatory module [43, 44] or the NF-κB module [36, 44] several feedback loops, positive and negative, are involved. Simple and typical negative feedback loops such as activation, by the transcription factor, of a gene coding for its own inhibitor provide homeostasis (steady state) for the cell. Negative feedback in conjunction with signal transmission delays may cause protein level oscillations in the cell. Positive feedback loops which amplify signal such as activation by the transcription factor of the gene coding for the inhibitor of the transcription factors inhibitor may introduce bistability into the system. Bistability allows the cell to make decisions such as whether to proceed to apoptosis or proliferation after DNA damage. An interesting fact is that positive feedback loops inside the cell usually work via a double negation in which the negative feedback is blocked. For example, positive feedback in p53 signaling pathway works through the blockade of the p53-inhibitor called Mdm2 by another transcriptionally p53-dependent molecule called PTEN (see [43]).

5.1.2 Experimental Data

Experimental data used for model fitting are frequently not quantitative. Results of the biological experiments are often presented in the form of Western-blot images (Fig. 5.1). These images present the level change of proteins marked by a fluorescent compound. These are only qualitative representations of the cell's response to stimuli. Even if transformed to numbers the results are in the arbitrary units. The knowledge we can obtain from these data is the response type such as oscillations or protein level increase after stimuli, and quantitative data such as oscillation period or time shifts but not the concentration or molecules number of a particular protein at a given time. We may estimate the fold change between the points on the Western-blot picture [32] remembering, however, that the result is approximate.

The Western-blot technique is based on the biological material obtained from a cell sample not from a single cell and therefore the results obtained should be considered as population level results. Despite simplicity and low cost, cell population experiments provide only results indicating the general population response but not the single cell response. In extreme cases, it is possible to imagine an experiment where each cell belongs to one of the two categories, one with high

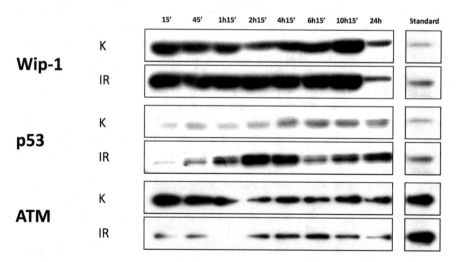

Fig. 5.1 Example of the Western-blot analysis result. The MCF-7 cancer cell line was irradiated with the dose of 10 Gy, then the protein level in the particular time points was checked (unpublished data communicated by Dr Patryk Janus)

level of the some protein and second with the low level. Population analysis will suggest the intermediate level of the protein which is of course false because in the population there is not a single cell with that level. Similarly, protein level oscillations in single cells, because of desynchronization among cells, will result in little or no oscillation observed in the population level experiments, which is also misleading.

Recently a large effort is placed in the field of biology to develop and improve the techniques of a single cell experiments. One can see the example of the results obtained by such method in the Geva-Zatorsky et al. article [19]. Geva-Zatorsky et al. used transfection to introduce into the cell specially modified p53 and Mdm2 gene copies. The modification consists in the attachment of the fluorescent particle at the end of the gene coding region. This allows real-time observation of the protein level change inside the single cell after a stimulus. This technique has, however, several problems that need to be resolved in the future.

First of all there is, similarly as in the Western-blot technique, a problem with converting the image information taken by using the microscope camera to the protein level change. This procedure has to be automated which is not trivial, taking into account that during the experiment cells could move, change shape, or divide. Currently most results are converted by a scientist or a technician who marks the fluorescent regions in the subsequent picture frames and then uses a computer program to convert fluorescent light intensity to numbers. This introduces significant noise to the final data which makes them difficult to interpret and to reproduce by another scientist.

The second and more important problem related to the presented technique is the gene copy number change in the cell. It is not always possible to remove the original gene copies from the nucleus before the transfection, sometimes it is to difficult to do or too expensive. Similarly, there is a problem how to determine the exact number of additional gene copies introduced or how to determine how many of them are really functional. This leads to the situation when a cell has more than the ordinary two functional gene copies. Proteins produced by the normal and additional genes play their usual roles in the cell but only those from extra copies are detectable by the fluorescent part detection. As a result, the level of the expression change might be misleading. Moreover, introduction of the additional gene copies increases the corresponding protein production rate compared to the unmodified cells. This results in a possible disturbance in the examined signaling pathway. The possibility arises that the observed cell response is specific to the modified cells and does not occur at all or occurs at different levels in the normal, unmodified cells, behavior of which we want to explore. It is therefore possible that, for example, the observed oscillation of the protein level after the stimulus is not the actual cell response to the stimulus but the methodology artifact [27].

However, despite all the caveats, there exist successful attempts to reconcile single-cell and cell-population dynamics even in large systems. One example is a model of the two arms of the innate immune response in a system imitating viral infection [8]. In this paper, single-cell fluctuations (including oscillations detected by Fourier analysis) have been shown to average-out to produce a coherent large-scale response, in the experiment and in the mathematical model.

5.1.3   Mathematical Modeling

There are several ways to mathematically describe the complex regulatory networks such as Direct Graphs (DG), Bayesian Networks (BN), Boolean Networks (BON), Generalized Logical Networks (GLN), Nonlinear Differential Equations (NLDE), Piecewise-Linear Differential Equations (PLDE), Partial Differential Equations (PDE), Stochastic Master Equations (SME), and Rule-Based Formalisms (RBF). These methods can provide static (DG, BN), or dynamic description, discrete (BN, BON, GLN, SME, RBF) or continuous (BN, NLDE, PLDE, PDE) in time or space. They can take into account the stochastic noises (BN, SME) or not, and provide qualitative (DG, BON, GLN, PLDE, RBF) or quantitative (BN, NLDE, PLDE, PDE, SME) results. They also vary in the simulation accuracy. Good description of the above methods formalism and their relative strengths and weakness can be found in [15]. Another method involves application of Petri Nets for deterministic, stoshastic and spatial models [6, 7]. Here we focus only on the most commonly used descriptions such as Direct Graph (DG), Ordinary Differential Equations (ODEs), and the Gillespie Algorithm (GA) originating from SME description of chemical reactions kinetic.

It is worth mentioning that in some cases stochastic description can be replaced by PDE with a generalized transport equation. For example, in [2] in which the authors describe structured cell population models in which the expected cell cycle time is a random variable from a gamma distribution dependent on the initial cell size. The cells that are larger at the beginning of life also have shorter life. The purpose is to stabilize the distribution of cell sizes in the population. Additionally the authors assume unequal cell division during which the daughter cell size is a random variable depending on the mother cell size. The proliferation process of such cell can be described stochastically as the branching process. The authors show that this description is equivalent to the deterministic description by a PDE of the generalized transport type, where transport is considered as the transport trough time.

5.1.3.1   Direct Graphs

The Direct Graphs are usually defined as tuple $G = (V,E)$ where V and E denote the set of vertices and the set of arcs, respectively, the latter defining relations between vertices. In the signaling pathways graph the vertex may represent a particular gene, mRNA, protein, or a state of the protein such as normal or phosphorylated form. The edge describes relations between them such as protein activation and deactivation, gene activation, production, degradation, or more complex such as formation or dissolution of complexes or enzymatic interaction.

The role of the direct graphs, in the presented approach, is to summarize the complex biological knowledge about the regulatory network structure and interactions in a simple, and therefore easy to understand, graph. Such representation can reveal dynamical structures such as negative and positive feedback loops or even whole subsystems of the complex system. An example of the direct graph description of the p53 signaling pathway, from [43], is given in Fig. 5.2.

The presented model contains 14 vertices which represent two forms of DNA (normal and damaged), mRNA of Mdm2 and PTEN and various forms of p53, PTEN, PIP, Akt, Mdm2 molecules. It also contains two special vertices described as "deg" (degradation) and "IR" (Ionizing Radiation) which are the inputs of this model.

We have two types of edges: solid, representing transitions (24 edges) and dotted, representing influences (11 edges).

Transitions include production of the p53 (mRNA stage for this protein is omitted) and mRNA of Mdm2 and PTEN; translations for Mdm2 and PTEN; activations and deactivations of PIP, Akt, Mdm2, and p53; Mdm2 transport from cytoplasm to the nucleus and DNA repair and damage processes. The last transition type is degradation which in the graph is shown as the edge between the given vertex and the vertex described as "deg."

Influences include enzymatic influences of PTEN on PIP, PIP3 on Akt and active Akt on Mdm2. They also represent p53 ubiquitination and therefore degradation by Mdm2, accelerated degradation of Mdm2 in the presence of damaged DNA and p53 driven Mdm2 and PTEN genes activation and DNA repair. The last influence presented on the graph is the DNA damage by IR.

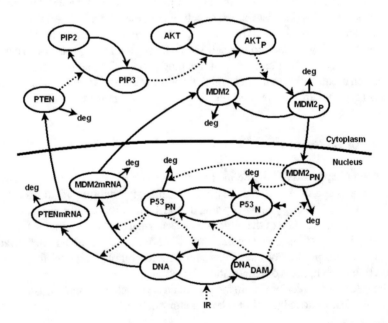

Fig. 5.2 Diagram of the p53 signaling pathway model proposed by Puszynski et al. in [43]

From the system dynamics point of view, direct graph presentation reveals two feedback loops in the p53 system. The first one, coupling p53 and Mdm2 is negative as p53 produces its own inhibitor. The second feedback is positive by double negation, as p53 produces PTEN which deactivates PIP. This leads to lack of Akt activation followed by lack of Mdm2 activation and by p53 inhibition.

5.1.3.2 Ordinary Differential Equations

ODEs constitute the most common mathematical formalism to mathematically describe a wide range of phenomena. In the signaling pathways description, variables describe the amount or the concentrations of genes, mRNA, or protein molecules. All reactions inside the cell are treated as if they were occurring in a well-mixed biochemical reactor. Accordingly, rules such as the mass action law apply. Therefore the right-hand side of the equations represent gains and losses of the considered species amounts resulting from all possible reactions in which these species are involved. This can be, for example, the gain representing protein production and loss representing its degradation. The basic chemical principles used in development of mathematical models of signaling pathway are presented in Sect. 5.1.4. Please notice that because of the nature of the biological system the variables must take only nonnegative values.

It seems worth mentioning that in a system of ODE with diagonal terms negative and off-diagonal nonnegative, positivity of solutions is preserved. For more details, please consult Murray's mathematical biology book [41].

To obtain the model solution which means the time courses of all variables one has to solve the system of ODE. Because signaling pathways description by ODEs usually leads to a system of strongly nonlinear equations, analytical solution is not possible and numerical simulations are required. The most commonly used numerical methods to solve the system of ODE are the Runge–Kutta methods. They differ in the order and coefficients. The general rule is that higher order methods are more accurate but also more time-consuming. Usually order 4 is accurate and fast enough for most application and therefore most used. To speed up the simulations one can use variable-step modification of the Runge–Kutta methods which usually combines fourth and fifth order or second and third order to determine the simulation step. More accurate description of the Runge–Kutta method with a description of its numerical implementation is given in Appendix D.

ODEs allow simple analysis of dynamical properties of the described system such as existence of the equilibrium points and determination of the attraction regions of this points, existence of the limit cycles, stability and controllability analysis. Various equilibrium points are often connected with various characteristic system behaviors such as proliferation or apoptosis. However, because as mentioned, real models of the signaling pathways are usually very complex and strongly nonlinear, analytical solution or even qualitative analysis is difficult or impossible to perform and therefore numerical analysis is the only practical solution.

Example of the ODE description of simple mathematical model is given in Sect. 5.1.4.4.

5.1.3.3   Gillespie Algorithm

In reality, all reactions which occur in the biochemical reactor result from stochastic events in which the molecules are created, degraded or react with each other. Probabilities of these events depend on many factors, such as the number of molecules involved in a particular reaction, temperature, enzyme concentrations, pH value, and so forth. When the number of reacting molecules is high, the deterministic approximation of the mathematical description of these stochastic events is accurate enough, but with small numbers of reacting molecules it is not (see Sect. 5.1.6 for further details). Therefore, stochastic description and following numerical simulation is required.

By the year 1976, when Gillespie published his work on the stochastic simulation of chemical reactions, the only effective approach to modeling of dynamics of stochastic systems was the so-called master equation. This approach requires construction of a system of ODE in which the variables are the probabilities that the system is in a given state at a given time. The state, in turn, is defined as a vector $S(t) = [S_1(t), \cdots S_N(t)]$, where $S_i(t)$ denote the number of molecules of type i at time t. Because every possible state of the system requires its own variable, the system of equations grows with the growing number of variables and their possible states. The solution of this system is demanding. Even if solved, it will not give the answer of the system evolution over time. Instead it answers the question how the particular probabilities changed over time. That answer does not provide a simple way to determine how many molecules of the particular peptide are inside the cell at a particular time, which is the basic question in the systems modeling.

Gillespie proposed to simulate only one of the possible system trajectories and then to use statistics to estimate the molecule concentration [20]. In the Gillespie approach stochastic simulation is used to describe the state vector $S(t)$ change over time, assuming that the initial state $S(0)$ of this vector is known. In each step of the Gillespie algorithm-based stochastic simulation, when the system is in a given time and state, two key decisions are made:

- when the next reaction will occur
- which of all possible reactions occurs at that time

Because both of these decisions are based on the randomly chosen numbers, the time evolutions received from the same initial conditions are different in each simulation run. It is worth noticing that each time the above decisions are made only one firing of the winning reaction occurs. This means that if we have N molecules A and M molecules B, and in the i-th reaction they form a complex AB, then if the i-th reaction wins, only one molecule of A and one molecule of B will form only one molecule of AB. For that reason Gillespie algorithm in its pure form is very time-consuming for complex systems. Therefore, during the years many modifications were proposed to speed up the simulations. Numerical implementation of the Gillespie algorithm and main modifications are discussed in Appendix D.

To run Gillespie algorithm-based simulation one has to determine propensities (probabilities per time unit) of all reactions possible in the simulated system. As in the ODE case, the basic chemical principles used in development of these propensities are presented in Sect. 5.1.4. An example of the system description by reaction propensities is given in Sect. 5.1.4.4.

An interesting modification of the original Gillespie algorithm, which couples it with ODEs, was provided by Haseltine and Rawlings [26]. They proposed to divide all reactions in the system into two subgroups: fast, which are generally the reactions with large number of reacting molecules, and slow, which are the reactions with only few reacting molecules. The fast reactions are modeled by ODEs, while slow ones by a pure Gillespie algorithm. Proposed modification, although challenging from the numerical implementation point of view, provides a fast, accurate stochastic simulations. Numerical implementation of the Haseltine–Rawlings modification is given in Appendix D. The example of the model defined according to this modification is given in Fig. 5.7

5.1.4 Basic Principles Used in Development of the Mathematical Model

Building mathematical models of signaling pathways is based on the rules known in biochemistry. Reacting proteins are in the biochemical terms, enzymes, substrates, and products. Below, the main laws and types of reaction used in the signaling pathway models building are described.

5.1.4.1 Mass Action Law

The mass action law is a fundamental law of the chemistry, biochemistry, and modeling of the signaling pathways. It was proposed in 1864 [55] by Guldberg and Waage and therefore it is often called Guldberg–Waage law. According to it, it is assumed that the probability of the event that two reacting molecules meet does not depend on their previous actions but only on their current amount in the given volume and mean kinetics energy of their movement. Then, the rate of the elementary chemical reaction is proportional to the effective concentrations of all participating reactants.

The above results from the fact that the reaction rate depends on the number of collisions occurring in the given time period (see also the introduction to the chemical reaction networks in the form of Martin Feinberg lectures is found on the Ohio State University web page [16]).

This leads to the following conclusions:

- if the concentration of any of the involved substrates is equal to zero, the reaction does not occur,
- increasing the concentration of substrates results in increased reaction rate while reducing the concentration of any substrate results in reduced reaction rate,

- if the reaction is reversible and it occurs in the closed system, then increasing reaction rate in one direction, for example by using the proper catalyst, results in the increased reaction rate in the opposite direction.

In the development of the mathematical models of the signaling pathways, two types of reactions, based on mass action law, are most commonly used: addition reaction and shift reaction. The addition reaction is a simple coupling reaction of several substrates in a single product. For example, two proteins A and B are combined in an AB complex. Assuming that reaction is irreversible, so the AB complex cannot dissolve to A and B proteins, we can present it using the following formula:

$$A + B \rightarrow AB. \tag{5.1}$$

The rate of the above reaction depends on the probability of the effective collisions of the A and B proteins. Assuming that this probability is proportional to the chemical activity of the reacting substrates we can present the above reaction rate as:

$$r = k \times A \times B, \tag{5.2}$$

where k is the reaction rate coefficient, and r is the so-called rate of the A and B substrates disappearance. Assuming that the complex AB does not undergo further changes, dissolution, or degradation, we can derive the ODE describing the rate of its formation:

$$\frac{d}{dt}AB(t) = k \times A(t) \times B(t). \tag{5.3}$$

Shift reaction is a reversible reaction, so the newly formed AB complex can dissolve back to the A and B proteins:

$$A + B \leftrightarrow AB, \tag{5.4}$$

which, under the assumption that the A and B proteins and AB complex are not involved in any other reactions, results in the following ODE describing the rate of the AB complex amount change:

$$\frac{d}{dt}AB(t) = k_a \times A(t) \times B(t) - k_d \times AB(t), \tag{5.5}$$

where k_a is a complex formation rate coefficient and k_d is a complex dissolution rate coefficient.

Additionally, first-order reactions, such as molecules degradation, activation and inactivation are usually described this way.

5.1.4.2 Michaelis–Menten Kinetics

In 1913 Leonor Michaelis and Maud Menten proposed a mathematical description of the catalytic (enzymatic) reactions, in which addition of an enzyme results in a huge acceleration of the substrate to product conversion [38]. Assuming that the product cannot be transformed back to the substrate and that the enzyme does not react with the product, the reaction has the following form:

$$E + S \leftrightarrow ES \rightarrow E + P \tag{5.6}$$

Michaelis and Menten proposed to describe the reaction velocity as a function of substrate concentration by the formula:

$$v_0 = v_{\max} \times \frac{S(t)}{k_M + S(t)} \tag{5.7}$$

The best derivation of the Michaelis–Menten formula was provided by Briggs and Haldane in 1925 [9]. It is as follows: denoting by k_1 the rate of formation of the enzyme-substrate complexes ES, by k_2 the rate of formation of the product P and enzyme E from the ES complex, and by d_1 the rate of dissolution of the ES complex into the enzyme E and substrate S, we can present the ES complex amount change rate as the following ODE:

$$\frac{d}{dt} ES(t) = k_1 \times E(t) \times S(t) - d_1 \times ES(t) - k_2 \times ES(t) \tag{5.8}$$

Under the assumptions that the ES complex concentration changes much slower than that of the product and substrate and following the quasi-steady state assumption that we have enough substrate to neglect its loss during the reaction and the enzyme amount during the reaction does not change, we obtain

$$\frac{d}{dt} ES(t) = 0 \tag{5.9}$$

$$E_{\text{tot}} = E(t) + ES(t) = \text{const}, \tag{5.10}$$

and therefore

$$0 = k_1 \times (E_{\text{tot}} - ES(t)) \times S(t) - (d_1 + k_2) \times ES(t), \tag{5.11}$$

which we can transform to

$$S(t) \times E_{\text{tot}} = S(t) \times ES(t) + ES(t) \times \frac{d_1 + k_2}{k_1}. \tag{5.12}$$

Introducing

$$kM = \frac{d_1 + k_2}{k_1}, \tag{5.13}$$

which we call the Michaelis–Menten constant, and transforming Eq. (5.12) to the form

$$S(t) \times E_{\text{tot}} = (kM + S(t)) \times ES(t) \tag{5.14}$$

we obtain

$$ES(t) = \frac{S(t) \times E_{\text{tot}}}{kM + S(t)}, \tag{5.15}$$

which after the substitution to the ODE of the product amount change rate

$$\frac{d}{dt}P(t) = k_2 \times ES(t) = v_0 \tag{5.16}$$

gives the equation of the product P formation rate v_0 as a function of the substrate S concentration, called the Michaelis–Menten equation:

$$v_0 = v_{\text{max}} \times \frac{S(t)}{kM + S(t)} \tag{5.17}$$

where $v_{\text{max}} = k_2 \times E_{\text{tot}}$ stands for the maximum possible reaction rate and kM can be interpreted as the substrate concentration at which the reaction reaches a half of its maximum rate.

5.1.4.3   Hill's Coefficient

In 1910 Archibald Vivian Hill noticed, during examination of the oxygen to hemoglobin binding process, that if many of the hemoglobin particles concentrate then the oxygen binding is more efficient [28]. Such dependence or reversed dependence in which the reaction rate is lower when the substrates of one kind concentrate is observed in many types of intracellular processes. Hill proposed the formula describing these dependencies. According to it, the amount of the binding sites Q occupied by the ligand is equal to

$$Q = \frac{L^n}{k^n + L^n} \tag{5.18}$$

where L stands for the ligand concentration and k is the ligand concentration at which half of the binding sites are occupied. The n parameter is called Hill's coefficient. The Hill's coefficient value equal to 1 means independent ligand binding. Value greater than 1 means positive ligand cooperation which results in a more efficient ligand binding in the presence of other ligands, while value smaller than 1 means negative cooperation which results in a less efficient ligand binding. Please notice that when $n = 1$ the Hill's equation is equivalent to the Michaelis–Menten equation.

5.1.4.4 Toy Models

To illustrate how to apply the above rules to build deterministic and stochastic models, let us consider two models: a simple auto-regulation model and the model of $p53$ and $Mdm2$ cross-regulation.

The simple deterministic auto-regulation model consists of three variables representing the number of active alleles (gene state), the number of mRNA molecules, and the number of protein molecules. The first equation describes gene states change over time:

$$\frac{d \; GENE}{dt} = q_a \cdot (N_{GENE} - GENE) - q_d \cdot GENE \cdot PROTEIN. \tag{5.19}$$

The first term stands for spontaneous gene activation, which depends on the total number of alleles N_{GENE}, while second for protein driven gene deactivation. The second equation describes mRNA production, whose rate depends on the number of active alleles and its spontaneous degradation:

$$\frac{d \; mRNA}{dt} = t_1 \cdot GENE - d_1 \cdot mRNA, . \tag{5.20}$$

The last equation describes protein production from mRNA and its spontaneous degradation:

$$\frac{d \; PROTEIN}{dt} = s_1 \cdot mRNA - d_2 \cdot PROTEIN. \tag{5.21}$$

Gene deactivation by the protein which is the product of this gene constitutes negative feedback loop.

Stochastic description of the above auto-regulation model requires definition of the propensities of all reactions involved in it. We have six of them: gene activation, gene deactivation, mRNA production, mRNA degradation, protein production, and protein degradation:

$$a_1 = q_a \cdot (N_{GENE} - GENE),$$
$$a_2 = q_d \cdot GENE \cdot PROTEIN,$$
$$a_3 = t_1 \cdot GENE,$$
$$a_4 = d_1 \cdot mRNA,$$
$$a_5 = s_1 \cdot mRNA,$$
$$a_6 = d_2 \cdot PROTEIN.$$

(5.22)

$p53$ and $Mdm2$ cross-regulation toy model is more complicated as it contains strongly nonlinear terms. It includes four types of molecules: $p53$, $Mdm2_c$, $Mdm2_n$, and $PTEN$ whose concentrations dynamics is governed by the following equations:

$$\frac{d\ p53}{dt} = p_1 - d_1 \cdot p53 \cdot (Mdm2_n)^2,$$

(5.23)

$$\frac{d\ Mdm2_c}{dt} = p_2 \frac{(p53)^4}{(p53)^4 + k_2^4} - k_1 \frac{k_3^2}{k_3^2 + (PTEN)^2} \ Mdm2_c - d_2 Mdm2_c,$$

(5.24)

$$\frac{d\ Mdm2_n}{dt} = k_1 \frac{k_3^2}{k_3^2 + (PTEN)^2} \ Mdm2_c - d_2 Mdm2_n,$$

(5.25)

$$\frac{d\ PTEN}{dt} = p_3 \frac{(p53)^4}{(p53)^4 + k_2^4} - d_3 \cdot PTEN.$$

(5.26)

The first equation describes $p53$ production and its degradation driven by the nuclear $Mdm2$. Notice the mass action law term describing degradation. It involves $p53$ double ubiquitination caused by $Mdm2_n$, which corresponds to the square in this term.

The second equation describes change of the cytoplasmic $Mdm2$ amount over time. The first term represents p53-dependent $Mdm2_c$ production. The second term represents its transport to the nucleus which is negatively regulated by $PTEN$. The last term represents spontaneous $Mdm2_c$ degradation. Please notice Hill terms with Hill coefficients equal to 4 in first and 2 in second term. The power 4 corresponds to the fact that transcription factor activity of $p53$ is provided by its tetramers. The square is due to the simplification of $PTEN$ positive feedback in the toy model compared to the full model (see Fig. 5.2). It compensates for the missing part of the dynamics due to this simplification.

The third equation describes the change of the nuclear fraction of $Mdm2$ over time. Based on mass conservation law the first term is analogical with the second term in the second equation as it describes nuclear transport of cytoplasmic $Mdm2$. The second term describes spontaneous degradation of $Mdm2_n$ whose rate is the same as that of $Mdm2_c$.

The last equation stands for $PTEN$ production driven by $p53$ tetramers and its spontaneous degradation.

Because of the strong nonlinearity of the *p53-Mdm2* model its presentation in the form of reaction propensities is not as easy as for auto-regulation model. The trivial transformation may lead to inaccuracies during the simulation. Because this nonlinearity usually results from simplification of multi-step reaction to a single term (e.g., Michaelis–Menten or Hill dynamics) it may be necessary to split the nonlinear term to simple step by step reactions and define propensities of each of this reaction. Taking into account that nonlinear terms usually describe fast reactions according to the Haseltine–Rawlings postulate, development of a hybrid deterministic-stochastic model may be the best strategy when stochastic simulations are required.

Please notice that presented toy model of *p53-Mdm2* cross-regulation similarly as the full model presented in Fig. 5.2 contains two feedback loops: negative coupling *p53* and *Mdm2*, and positive through *PTEN* as its block *Mdm2* nuclear entry. The first, auto-regulation model also contains negative feedback loop as protein deactivates its own gene. One can notice, however, that identification of dynamical structures is more difficult when the model is presented in terms of the ODE or reaction propensities compared to the graph presentation.

For further explanation of the p53 signaling pathway modeling, see [43].

5.1.5 Stochastic or Deterministic Model?

A living cell may be considered a complex biochemical reactor in which the protein molecules react with each other according to the specific reaction rules. As mentioned, the rate of these reactions depends on the concentrations of the reacting molecules and environment conditions such as the temperature or pressure, which all are condensed to single reaction coefficients. In general every single reaction such as the binding of the two substrate molecules to form a single complex molecule should be considered a stochastic event. The probability of occurrence of a single event and hence the exact time when the reaction occurs depends on the parameters. Moreover, a single stochastic event such as single gene activation may result in a hundred of thousand of the corresponding protein molecules, so the impact of this single event may be crucial for the cell fate. The above argues for the stochastic models.

On the other hand, stochastic description of complex systems, such as the signaling pathways, may be difficult and the resulting model difficult to analyze. Reference to the concepts known from the deterministic models analysis such as equilibrium points or limit cycles may be problematic. For these reasons it may be useful to develop the deterministic model of the considered biological system, based on the ODEs. In addition bifurcation analysis, which may be used, for example, to determine the models robustness to the change of parameter value [27] is easier for deterministic description than stochastic. Moreover, the deterministic model simulations are faster than the stochastic ones because of a smaller number of the computational steps. Therefore, deterministic models are often use as an approximation of an average cell behavior (Fig. 5.3).

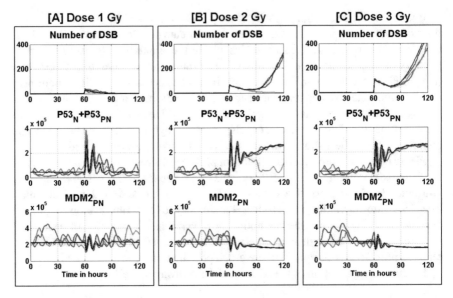

Fig. 5.3 Example of the deterministic and stochastic results of the p53 model simulation [43]. The *black line* symbolizes the deterministic result while *color ones* for the single stochastic realizations. Please notice that for the dose of 2 Gy deterministic approximation corresponds to an apoptotic solution (high level of the $P53_N + P53_{PN}$ at the end) while one of the stochastic realizations (*brown*) corresponds to a survival solution

Although deterministic models are easier to analyze and require less computational effort to simulate than stochastic ones, their averaging nature can be their major flaw. As one may notice in biological experiments such as in the Geva-Zatorsky experiment [19], individual cells of the same type differ substantially in their response to exactly the same stimulus. As a result, the cell's response to the DNA damage after irradiation may take the form of the programmed cell death, called apoptosis, or cells proliferation. Deterministic description by its nature does not allow for a differentiated response, assuming the same initial conditions, parameters, and control signals. As shown before, even interpretation of the deterministic results in the terms of "mean" cell behavior may be misleading because, as a result of the cells response desynchronization, there may not be a single cell in the population which follows this mean behavior.

Therefore if our goal is to properly describe the response of a single cell, from the population exhibiting large variation in response to a stimulus, stochastic modeling is necessary. However if we want to describe the response of the homogeneous cells population or analyze our model by using methods designed for the deterministic models analysis, we should build such models or develop the deterministic approximation of the existing stochastic models.

5.1.6 Deterministic Approximation of the Stochastic System

The easiest approach to the deterministic approximation is through replacement of the reaction rules of the individual reactions by the ODE describing the time course of variables of interest (concentrations of chemical species).

Let us consider mRNA production by an active gene. The number of mRNA molecules in the time interval $(t, t + \Delta t)$ increases when production occurs or decreases when degradation occurs. Let us assume that at time $t + \Delta t$ system is in state n which means we have exactly n mRNA molecules. This state can be reached by production if the system was in the state $n - 1$ at time t or by degradation if the system was in the state $n + 1$ at time t. The system can leave the state n when production occurs, going to $n + 1$, or degradation occurs, going to $n - 1$, as shown in Fig. 5.4.

Production rate of the mRNA does not depend on the state and is equal to s; degradation rate depends on the state and is the product of the known coefficient d and the number of mRNA molecules in the given state, for example dn for state n. Denoting by x_n the probability that at time t the system is in state n we can write its master equation:

$$\frac{d}{dt}x_n = s x_{n-1} + d x_{n+1} (n + 1) - s x_n - d x_n n. \qquad n \geq 0 \qquad (5.27)$$

where $x_{-1} = 0$ by definition.

Stationary probability distribution of the system (5.27) is the Poisson distribution with mean equal $\lambda = s/d$:

$$P_n = \frac{e^{-\lambda} \lambda^n}{n!} \qquad (5.28)$$

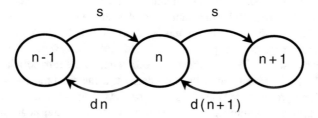

Fig. 5.4 Illustration of the mRNA production/degradation system changes. Notation: n, number of the mRNA molecules; s, production rate; d, degradation rate

In the Poisson distribution the mean is equal to the variance, so the coefficient of variation ("noise") σ defined as the ratio of the standard deviation to the mean is given by

$$\sigma = \frac{\sqrt{\lambda}}{\lambda} = \frac{1}{\sqrt{\lambda}}. \tag{5.29}$$

According to Eq. (5.29) the noise decreases with the increase of the mean amount of the molecules, so if the mean is high enough the deterministic (mean) approximation, describing the rate of the change of amount of mRNA molecules in time, is given by

$$\frac{d}{dt}n = s - d \cdot n, \tag{5.30}$$

and it is justified by the small error it entails. Please notice that in the steady state the amount of the mRNA molecules n is equal to the Poisson mean λ.

Stationary probability density for more complex reactions may not be given by the Poisson distribution but by other distributions. However, we may assume that also in these distributions noise decreases with the increase of the mean number of the reacting molecules. Because the number of the mRNA or protein molecules in the signaling pathways models is much higher than 1 such deterministic approximation is rational.

5.2 Intracellular Processes as Objects of Control

There is a significant number of diseases with the genetic background such as cancer. Until now despite a significant effort of the scientists around the world there is no solution to the cancer problem. Even if we know that the cell populations are heterogeneous and organisms are not identical, there is no personalized therapy. Our knowledge of the intracellular mechanisms of cancer is not complete as well. Until now the cancer problem was considered on the organism, tissue or cell level and a set of hallmarks of cancer was proposed [24] and [25]. The new promising approach to the problem is to consider the cancer as dysfunction of the intracellular regulation processes. In this approach the cell is considered as a chemical reactor where genes, transcripts, and proteins are substrates and products. Their mutual interactions and dependencies form a complex system with many positive and negative feedback loops. Mathematical description of this system allows applying control engineering and systems theory methods to the system analysis. In the proposed approach the cancer is triggered by a specific change of model parameters. This causes the change of the types of equilibrium points and/or shifts of the bifurcation points compared to the normal (healthy cells) state. In contrast to mechanical elements, a biological system cannot be repaired physically. However, external control signals can be used to bypass the defective part of the signaling pathway and restore desired cells behavior (like the death of the cancer cells). In the recent years we have observed a growing number of publications concerning the technical possibilities of artificial molecules inserted into the cell which can be considered control signals [13, 49, 53]. Some of them are based on interfering RNA [13], while others on chemical compounds [53]. From the control-engineering point of view this means that the role of control is to return the equilibrium points and/or bifurcation points to the original state.

Application of the control signals in clinics requires considering pharmacokinetics and pharmacodynamics of the drug [12]. As the drug is usually delivered by oral application or injection it has to be transported from the stomach, bowel or the injection place to the organ affected by cancer. The cells or whole body influence on the drug such as transport between organs and plasma, transport through plasma, or drug removal is called pharmacokinetics. The drug influence on cells such as DNA damage, complex creation disturbance, or microtubules destruction is called pharmacodynamics. The circadian rhythm, cell cycle as well as stochastic gene-switching has to be also considered when planing real therapies as they influence drug pharmacodynamics thus affecting the therapy results. The basic models of pharmacokinetic and pharmacodynamics of the drugs were described in Chap. 2.

From system engineering point of view, the disease state may be represented as a displacement of the bifurcation point or change of the equilibrium point type. Such approach may lead to new therapy protocols, which are not currently. As shown in Fig. 5.5, the proposed methodology can also be used for personalized therapy.

To summarize, the proposed system engineering approach to the anticancer therapies is as follows:

1. identification of the structure and parameters of the control object (intracellular signaling pathway). In the personalized therapy they would be patient-dependent,
2. development and computational simulation of the deterministic and stochastic models of the object of control,
3. analysis of the differences in the response of the normal (healthy cells) and damaged (cancer cells), focused on equilibrium points, bifurcation and sensitivity analysis,
4. identification of potential targets for the control signal designed to restore normal behavior of control objects,
5. development of the optimal control protocol which will minimize the doses and side effects (e.g., number of the dead healthy cells) and maximize therapeutic effects (e.g., increase of the apoptotic fraction of the cancer cells),
6. experimental verification of the optimal control protocol developed at the experimental stage and application of the optimal control protocol in personalized therapy.

In subsequent sections we present an example of stability and bifurcation analysis applied to a model of a signaling pathway. Further details of the bifurcation analysis

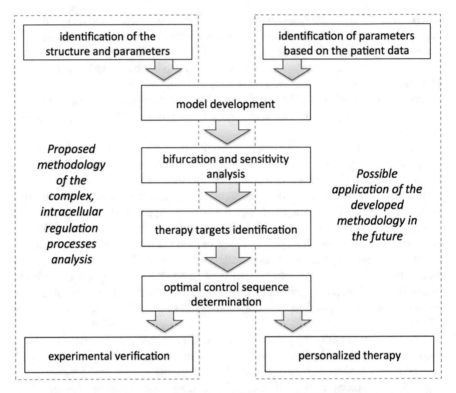

Fig. 5.5 An approach to personalized therapy based on systems biology methods

can be found in Appendix C while further details of the equilibrium points and their stability analysis can be found in handbooks of systems dynamics such as [29].

Determination of the number and types of equilibrium points is one of the most fundamental steps in analysis of dynamical systems. It provides a glimpse into a qualitative responses of the system as well as constitutes the first step into more advanced investigation of model properties, including bifurcation analysis. In this section an illustrative example of analysis of equilibrium points in a signaling pathway model is introduced (for definitions and basic methods, see Appendix A).

One of the most important signaling pathways related to carcinogenesis is the p53 signaling pathway. It controls cell responses that include cell cycle arrest, DNA repair, apoptosis, and cellular senescence to input signals, such as DNA damage, oncogene activation, heat and cold shock, and others [43]. The p53 protein production and functionality is damaged at some point in 50% of the known cancers, in the rest of them other proteins in the p53 regulatory units do not work properly. Therefore there is a huge effort around the globe to investigate the p53 signaling pathway dysfunctionality and find the ways to repair it.

There are many feedback loops in the p53 signaling pathway. Among them one negative (coupling p53 with Mdm2) and one positive (coupling p53 with PTEN,

PIP3, Akt, and Mdm2) are crucial for the proper system behavior. Existence of the negative feedback assures homeostasis of healthy cells and oscillatory responses of DNA-damaged cells, which are persistent when DNA repair is inefficient and the positive feedback loop is broken (like in MCF7 cells). The positive feedback blocks inhibitory actions of Mdm2 on p53 by sequestering most of Mdm2 in cytoplasm, so it may not longer prime the nuclear p53 for degradation. This positive feedback loop works as a clock: It gives the cell some time for DNA repair, but when the repair is inefficient, it makes the active p53 to rise to a high level and triggers transcription of proapoptotic genes. As a result, small DNA damage may be repaired and cell may return to its initial "healthy" state, while the extended damage results in apoptosis.

From the system engineering point of view, in normal healthy cells we should have two attractors as long as the p53 dynamics is considered: one, when there is no DNA damage and the p53 level is low, and second when the DNA damage is too extensive to be repaired, and p53 levels rise above the apoptotic threshold. Between these attractors there exists a region of the damped oscillations during which the cell tries to repair its DNA (Fig. 5.6a). Certain mutations, which lead to cancer, introduce bifurcations, the effect of which may be, for example, blocking

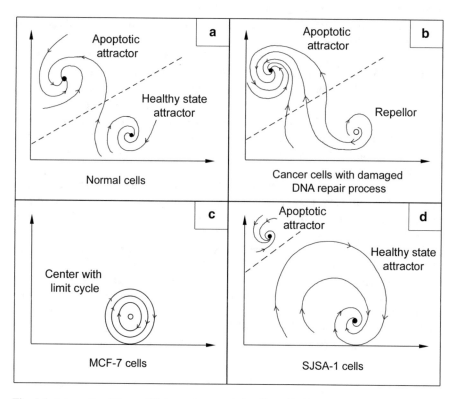

Fig. 5.6 Schematic of the equilibrium point-type in healthy cells (**a**) and some of their changes in cancer (**b**)–(**d**)

the DNA repair ability of the cell. In that case the first attractor representing the no DNA damage state becomes a repellor. That means that even the smallest DNA damage cannot be repaired and in time the positive feedback loop pushes the system to the apoptotic attractor (Fig. 5.6b). Another possibility is that as in MCF-7 cancer cells not only DNA damage repair processes, but also the positive feedback will be damaged. In that case we still have first repellor but this time it is the center of the stable limit cycle and the apoptotic attractor disappears. The reason is the action of the negative feedback loop, which makes the p53 and Mdm2 levels oscillate continuously (Fig. 5.6c). The cell damage which lead to cancer states besides abrogation of certain processes can also have the form of amplification or dampening of the certain processes. For example in SJSA-1 cells, because of the chromosome mutation, we can observe MDM2 overexpression. This does not change the equilibrium point types. We still have two attractors and a dampened oscillation between them but in this case it will be more difficult to switch the system to the apoptotic state (Fig. 5.6d).

Thus, system engineering approach not only allows us to determine the changes in the equilibrium point types but also, through a careful systems dynamics and sensitivity analysis, to propose a proper control to restore the cells normal behavior, which could be death of the cancer cells with irreparable DNA.

Bifurcation type analysis is useful for model validation and identification of the differences between normal and cancer cells. Model validation is based on the bifurcation existence in general and sensitivity of the oscillations period and amplitude. Knowing the biological systems behavior from the experiments we can check if the model behavior, for example the amplitude or period conservation is qualitatively similar. In general, systems based on a super Hopf bifurcation (see Appendix C) have oscillations with a relatively well-conserved period (it changes slowly with the bifurcation parameter), but a varied amplitude. In contrast, for the SNIC and SL bifurcations amplitude is fixed, but the period of oscillations may be arbitrarily high.

The normal and cancer cells may differ with respect to the type of the bifurcation and/or in the bifurcation point localization. This may be the cause of cancer, an example provided by 25-fold amplification of the expression of the Mdm2 gene in the SJSA-1 cancer cells.

The differences between the normal and cancer cells are related to the change of system and model parameters reflected by the eigenvalues of the Jacobian matrix. Knowing the difference, we can check for the possibilities of reversing such changes of eigenvalues types and values by modifying parameters of the mathematical model. This can be achieved, for example, by introduction of an additional control signal such as interfering RNA or chemical particles, the role of which is modification of the reaction rates. Examples follow in subsequent sections.

5.3 Example of the Control Signals in the Signaling Pathways

One of the basic problems concerning simulation of the intercellular signaling pathways is the appropriate definition of the controlling signal which could be successfully introduced in vivo in order to control cells response based on the designed mathematical model.

Analysis of equilibrium points and bifurcation provides information of the cancer cell difference with respect to the normal cell. Sensitivity analysis provides information about best targets to hit. The next step in the systems engineering approach to the anticancer therapies is to combine the determined targets with the biological knowledge of the possible ways to hit the targets. In other words, we have the targets ranking and now we have to determine which targets are reachable by known therapies.

Sensitivity analysis will give us the ranking of parameters related to the gene copies number, transcription and translation, degradation, complexes formation, and activation and deactivation parameters. Although in vitro it is possible to change the number of gene copies or block the transcription by artificial modification of gene promoter region in single cells, it is not possible in vivo and therefore it cannot, at least for now, serve as the basis of anticancer therapies. It is possible, however, to change the number of proteins targeting the corresponding mRNA. More precisely we can use the iRNA — interfering RNA (siRNA or miRNA) to change the mRNA degradation rate or block its entrance to the ribosomes [13]. Degradation control of mature proteins is not so easy. Usually it requires attaching at least four ubiquitin particles to the target peptide or disturbing the target protein deubiquitilation processes. The problem is that usually these processes are not controlled by specifically aimed molecules so it is almost impossible to control the processes at the target molecule only. A better idea is to influence the complex formation rates such as formation of enzyme-substrate complexes and therefore protein phosphorylation, acetylation, or ubiquitilation. This can be done by using proper chemical compounds such as Nutlin-3 which influences p53-Mdm2 complex formation reaction.

Control signals such as the interfering RNA and chemical particles (Nutlin-3) are discussed below. To show the practical implementation of these controls we will use as an example the p53 signaling pathway.

5.3.1 p53 Signaling Pathway

In Fig. 5.7 we present equations of the hybrid stochastic-deterministic model of the p53-Mdm2 signaling pathway. Exact description of the presented model can be found in [43]. In the normal cells without DNA damage the p53 is maintained at a low level by the action of a negative feedback loop consisting of p53 and its

Stochastic part: reaction probabilities:

$$P^b(t, \Delta t) = \Delta t \, (q_0 + q_1 \, P53^2_{np}(t))$$

$$P^d(t, \Delta t) = \Delta t \, q_2$$

$$P^{DAM}(t, \Delta t) = \Delta t \, d_{DAM} \, R$$

$$P^{REP}(t, \Delta t) = \Delta t \, \frac{d_{REP} \, P_A(t)}{N(t) + N_{SAT} \, P_A(t)} \, N(t)$$

Deterministic part: ODEs:

$$\frac{d}{dt} PTEN(t) = t_1 \, PTEN_t(t) - d_2 \, PTEN(t)$$

$$\frac{d}{dt} PIP_p(t) = a_2 \, (PIP_{tot} - PIP_p(t)) - c_0 \, PTEN(t) \, PIP_p(t)$$

$$\frac{d}{dt} AKT_p(t) = a_3 \, (AKT_{tot} - AKT_p(t)) \, PIP_p(t) - c_1 \, AKT_p(t)$$

$$\frac{d}{dt} MDM(t) = t_0 \, MDM_t(t) + c_2 \, MDM_p(t) - a_4 \, MDM(t) \, AKT_p(t)$$

$$- \left(d_0 + d_1 \, \frac{N^2(t)}{h_0^2 + N^2(t)} \right) \, MDM(t)$$

$$\frac{d}{dt} MDM_p(t) = a_4 \, MDM(t) \, AKT_p(t) - c_2 \, MDM_p(t) - i_0 \, MDM_p(t)$$

$$+ e_0 \, MDM_{pn}(t) - \left(d_0 + d_1 \, \frac{N^2(t)}{h_0^2 + N^2(t)} \right) \, MDM_p(t)$$

$$\frac{d}{dt} MDM_{pn}(t) = i_0 \, MDM_p(t) - e_0 \, MDM_{pn}(t) - \left(d_0 + d_1 \, \frac{N^2(t)}{h_0^2 + N^2(t)} \right) \, MDM_{pn}(t)$$

$$\frac{d}{dt} P53_n(t) = p_0 - \left(a_0 + a_1 \, \frac{N^2(t)}{h_0^2 + N^2(t)} \right) \, P53_n(t) + c_3 \, P53_{pn}(t) - (d_3 + d_4 \, MDM^2_{pn}(t)) \, P53_n(t)$$

$$\frac{d}{dt} P53_{pn}(t) = \left(a_0 + a_1 \, \frac{N^2(t)}{h_0^2 + N^2(t)} \right) \, P53_n(t) - c_3 \, P53_{pn}(t) - (d_5 + d_6 \, MDM^2_{pn}(t)) \, P53_{pn}(t)$$

$$\frac{d}{dt} MDM_t(t) = s_0 \, (G_{M1} + G_{M2}) - d_7 \, MDM_t(t)$$

$$\frac{d}{dt} PTEN_t(t) = s_1 \, (G_{P1} + G_{P2}) - d_8 \, PTEN_t(t)$$

Fig. 5.7 Hybrid, stochastic-deterministic model of the p53-Mdm2 signaling pathway from Fig. 5.2 published in [43]

inhibitor Mdm2 transcriptionally dependent on p53 (Fig. 5.2). After DNA damage occurs, Mdm2 degradation rate increases and p53 is phosphorylated and therefore stabilized. p53 plays the role of a transcription factor for many proteins related to cell cycle arrest and DNA repair as well as for Mdm2 and PTEN. Newly synthesized Mdm2 after its activation in cytoplasm by active Akt translocates to nucleus and degrades p53. This mechanism is responsible for p53-Mdm2 level oscillation. Meanwhile PTEN level rises and after some time PIP3 and active Akt level drops. As a result, if the cell was unable to repair damage up to this time, Mdm2 is trapped in the cytoplasm and nuclear level of p53 rises above apoptotic threshold. Existence of this threshold was recently confirmed experimentally by Kracikova [34].

Around 50 % of the known cancers have mutated p53 protein while in remaining ones p53 regulatory module is impaired. For example, in MCF-7 (breast cancer) cells PTEN gene is methylated and therefore not expressed [18, 35] so the positive feedback is broken, while SJSA-1 osteosarcoma cells have the 25-fold amplification of the Mdm2 gene [42] which gives the strong amplification of the negative feedback loop. Both mentioned p53 regulatory unit modifications result in p53 inability to reach the apoptotic threshold and cancer development.

From the system engineering point of view, Mdm2 gene amplification causes a significant reduction of the impact area of the apoptotic attractor (Fig. 5.6d) and a shift of the bifurcation points when DNA damage level is taken as a bifurcation parameter. This leads to SJSA-1 resistance to the ionizing radiation. Therapy of the osteosarcoma may be based on p53 level elevation, for example, by Mdm2 level reduction. Therefore taking p53 level as a model [43] output in global sensitivity analysis based on Sobol's indices (see next chapter for further details) for y_k subsets of the $k \in (1, 2, 3)$, (so a maximum of three parameters are changed at the same time) and excluding gene copy number and transcription parameters as possible targets, we receive the following parameter ranking:

1. Mdm2 transcript degradation rate
2. translation rate for Mdm2
3. p53 production rate
4. Mdm2-dependent p53 degradation rate
5. rate of Mdm2 activation by Akt

Confronting the above list with the biological knowledge we find that we can raise the p53 level in the Mdm2 overexpressed cells by lowering Mdm2 protein level using iRNA or disturbing p53-Mdm2 complexes creation, e.g., by using Nutlin-3.

5.3.2   Interfering RNA

Small RNAs (iRNA) regulate almost all aspects of cell life, including development, growth, differentiation, proliferation, apoptosis, stress response, cell metabolism, and cell signaling. The class of small RNAs includes small interfering RNAs (siRNA), microRNA (miRNA), and PIWIinteracting RNAs (piRNAs).

Small RNA incorporates two main mechanisms involving either miRNA or siRNA particles which become a part of a multiprotein complex called RISC responsible for the gene silencing processes. Unlike miRNA, siRNA do not naturally occur in human cells but their high specificity and the ability to almost entirely silence the expression level of a single target gene has proven its usefulness as a controlling agent in therapeutic studies [10].

miRNAs are known to be involved in many genetic disorders. Alterations in their expression profile were shown to be connected with a wide range of diseases, mainly cancer [51]. It was also shown that up- or down-regulation of different miRNAs may lead to modified sensitivity to chemo- and radio-therapy [30]. The natural mechanisms of miRNA-based gene expression regulation involve translation inhibition, mRNA destabilization, or both.

In vivo research adds a new group of problems, concerning iRNA transport into the cell. High efficiency of delivery mechanisms is one of the crucial elements of small RNA studies. siRNA/miRNA can be introduced to the cells in its mature form directly or with the help of transporting agents such as polymer-based nanoparticles. High efficiency of naked siRNA delivery was confirmed both with local and systematic delivery in animal models [33] although the usage of targeting agents can significantly increase the effectiveness of small RNA [50]. Another group of methods involves transfection of small RNA-expressing constructs with the help of either viral or a plasmid vectors which results in their production inside of the cell inducing long-term gene silencing effect [37].

Over the last few years the amount of siRNA-based therapeutic studies has significantly increased allowing to control the proliferation and apoptosis of specific cells based on the previously identified mathematical models describing interactions between various elements involved in intercellular signal transduction systems [4, 21, 45]. Mathematical description of small RNA has proven its usefulness providing precious pharmacodynamic simulation data used in various clinical trials including the design of CALAA-01 dosage system successfully tested on patients with solid cancer in a phase I clinical trial [14].

The largest disadvantages of siRNA mediated gene silencing include the fact that only negative regulation is available through the direct influence of siRNA. Positive regulation can be achieved only indirectly, through the blockage of target genes repressor. The problem does not exist in miRNA regulation, because gene silencing is related to overexpression of miRNA and overexpression of target mRNA is caused by knockdown of miRNA. Additionally aside from the in vitro studies it is extremely difficult to achieve a very high silencing effect in vivo due to transporting limitations and toxicity of high siRNA doses which also need to be addressed when constructing mathematic models.

A great majority of existing small RNA models do not consider the delivery mechanism [22, 47], are not applicable to mammalian cells [5], or are not supported by experimental studies [1], but still they provide valuable information that may aid the understanding of small RNA mechanism on molecular level.

5.3.3 Nutlin-3 as an Example of the Chemical Particle-Based Control

Mdm2 is a protein which serves as a very specific negative regulator of the p53. By incorporating the mechanism of molecular binding it targets p53 transactivation domain leading to its very efficient repression [23, 39]. Mdm2 uses three mechanisms which lead to the p53 blockage—by binding to it and impairing its ability to activate transcription, by favoring its nuclear export, and by enhancing its proteosomal degradation [11]. Increased Mdm2 expression was observed in many tumors resulting in p53 inhibition [17, 40], which in turn creates highly unstable conditions under which p53 loses its function as an efficient mechanism of DNA repair activation and damage dependent apoptotic switch, leading to various genetic disorders [54].

Nutlin-3 is a cis imidazoline derivative which prevents the Mdm2-p53 interactions. Nutlin-3 binds to the p53 pocket located on the surface of the Mdm2, preventing p53 from binding and therefore leading to p53-dependent pathway stabilization in defective Mdm2 rich, cell conditions [53]. Since it acts on the p53 regulatory module indirectly by the Mdm2 suppressor it loses its high functional efficiency in p53 deficient cells. The most beneficial effects of Nutlin-3 were observed in cells highly overexpressing Mdm2, although it may also increase p53module efficiency in tumors with normal Mdm2 expression [52].

Previous studies have shown that p53-MDM2 interaction blockage by the use of Nutlin-3 or other macromolecular approaches such as MI-219 [3] can lead to p53 activation and tumor growth inhibition becoming a novel therapeutic strategy [11, 48]. Apart from its potential clinical relevance which was already observed in vivo [52, 53], Nutlin-3 may serve as a molecular tool for the study of p53 signaling pathway in cells showing various types of aberrations in stress response mechanisms.

Nutlin-3 and its derivatives are currently very extensively evaluated since they can lead to stabilization and accumulation of p53, activation of p53-dependent genes and consequently leading to cell cycle arrest and/or apoptosis. Besides biological study also a simulation study of Nutlin-3 pharmacodynamics model is conducted for better understanding of its complex influence on the p53 signaling pathway [46]. Sample results given by the model are presented in Fig. 5.8. The largest advantage of Nutlin-3 is its non-genotoxic p53 activation without introducing DNA damage or post-translational p53 modifications making it a viable alternative to current cytotoxic chemotherapy drugs. Despite its ability to restore p53 in wild-type p53 tumor cells, high Nutlin-3 concentrations were also shown to inhibit cell proliferations even in cells lacking p53 [48].

Several investigations concern Nutlin's effects used in combination with other therapeutic agents, on various cells lines, in terms of apoptosis and cell cycle arrest. In a combination with ionizing radiation Nutlin can potentially increase lethality to cancer cells by maintaining a relatively small radiation dose [31], although this aspect still requires extensive research concerning the dosing schedule and molecular mechanisms leading to such actions.

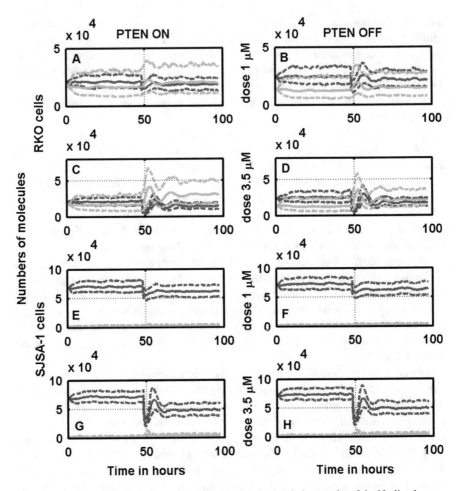

Fig. 5.8 Example of the hybrid, stochastic-deterministic simulation results of the Nutlin pharmacodynamics model published in [46]. *Solid lines* represent median level of phosphorylated p53 (*green*) or nuclear Mdm2 (*red*) of 500 cells, while *dashed lines* its 1st and 3rd quartiles. Specific dose of Nutlin was induced at time $t = 48\,\text{h}$ to the RKO or SJSA-1 cells with normal or silenced PTEN gene

References

1. J.C. Arciero, T.L. Jackson, D.E. Kirschner, A mathematical model of tumor-immune evasion and siRNA treatment pubmed. Discrete Contin. Dyn. Syst. **4**, 39–58 (2004)
2. O. Arino, M. Kimmel, Comparison of approaches to modeling of cell population dynamics. SIAM J. Appl. Math. **53**(5), 1480–1504 (1993)
3. A.S. Azmi, P.A. Philip, F.W. Beck, Z. Wang, S. Banerjee, S. Wang, D. Yang, F.H. Sarkar, R.M. Mohammad, MI-219-zinc combination: a new paradigm in MDM2 inhibitor-based therapy. Oncogene **30**, 117–126 (2011)

4. D.W. Bartlett, M.E. Davis, Insights into the kinetics of siRNA-mediated gene silencing from live-cell and live-animal bioluminescent imaging. Nucleic Acids Res. **34**, 322–333 (2006)
5. C.T. Bergstrom, E. McKittrick, R. Antia, Mathematical models of RNA silencing: unidirectional amplification limits accidental self-directed reactions. Proc. Natl. Acad. Sci. U.S.A. **100**, 11511–11516 (2003)
6. R. Bertolusso, M. Kimmel, bioPN: Simulation of deterministic and stochastic biochemical reaction networks using Petri Nets (2014). Software available at https://cran.r-project.org/web/packages/bioPN/index.html
7. R. Bertolusso, M. Kimmel, sbioPN: sbioPN: Simulation of deterministic and stochastic spatial biochemical reaction networks using Petri Nets (2014). Software available at https://cran.r-project.org/web/packages/sbioPN/index.html
8. R. Bertolusso, B. Tian, Y. Zhao, L. Vergara, A. Sabree, M. Iwanaszko, T. Lipniacki, A.R. Brasier, M. Kimmel, Dynamic cross talk model of the epithelial innate immune response to double-stranded rna stimulation: coordinated dynamics emerging from cell-level noise. PLoS One **9**(7), e103019 (2014). doi:10.1371/journal.pone.0103019
9. G.E. Briggs, J.B. Haldane, A note on the kinetics of enzyme action. Biochem. J. **19**, 338–339 (1925)
10. D. Bumcrot, M. Manoharan, V. Koteliansky, D.W. Sah, RNAi therapeutics: a potential new class of pharmaceutical drugs. Nat. Chem. Biol. **2**, 711–719 (2006)
11. P. Chene, Inhibiting the p53-MDM2 interaction: an important target for cancer therapy. Nat. Rev. Cancer **3**, 102–109 (2003)
12. M. Collins, R. Dedrick, Pharmacokinematics of anticancer drugs, in *Pharmacologic Principles of Cancer Treatment*, ed. by B.A. Chabner (Saunders, Philadelphia, 1982), pp. 77–99
13. M.E. Davis, The first targeted delivery of siRNA in humans via a self-assembling, cyclodextrin polymer-based nanoparticle: from concept to clinic. Mol. Pharm. **6**, 659–668 (2009)
14. M.E. Davis, J.E. Zuckerman, C.H. Choi, D. Seligson, A. Tolcher, C.A. Alabi, Y. Yen, J.D. Heidel, A. Ribas, Evidence of RNAi in humans from systemically administered siRNA via targeted nanoparticles. Nature **464**, 1067–1070 (2010)
15. H. De Jong, Modeling and simulation of genetic regulatory systems: a literature review. J. Comput. Biol. **9**(1), 67–103 (2002)
16. M. Feinberg, *Lectures on Chemical Reaction Networks* (1979). http://www.crnt.osu.edu/LecturesOnReactionNetworks
17. D.A. Freedman, L. Wu, A.J. Levine, Functions of the MDM2 oncoprotein. Cell Mol. Life Sci. **55**, 96–107 (1999)
18. J.M. García, J. Silva, C. Peña, V. Garcia, R. Rodriguez, M.A. Cruz, B. Cantos, M. Provencio, P. España, F. Bonilla, Promoter methylation of the PTEN gene is a common molecular change in breast cancer. Genes Chromosom. Cancer **41**, 117–124 (2004)
19. N. Geva-Zatorsky, N. Rosenfeld, S. Itzkovitz, R. Milo, A. Sigal, E. Dekel, T. Yarnitzky, Y. Liron, P. Polak, G. Lahav, U. Alon, Oscillations and variability in the p53 system. Mol. Syst. Biol. **2**, 2006.0033 (msb4100068) (2006)
20. D.T. Gillespie, A general method for numerically simulating the stochastic time evolution of coupled chemical reactions. J. Comput. Phys. **22**, 403–434 (1976)
21. M.A. Groenenboom, P. Hogeweg, The dynamics and efficacy of antiviral RNA silencing: a model study. BMC Syst. Biol. **2**, 28 (2008)
22. M.A. Groenenboom, A.F. Maree, P. Hogeweg, The RNA silencing pathway: the bits and pieces that matter. PLoS Comput. Biol. **1**, 155–165 (2005)
23. P. Hainaut, M. Hollstein, p53 and human cancer: the first ten thousand mutations. Adv. Cancer Res. **77**, 81–137 (2000)
24. D. Hanahan, R.A. Weinberg, The hallmarks of cancer. Cell **100**(1), 57–70 (2000)
25. D. Hanahan, R.A. Weinberg, Hallmarks of cancer: the next generation. Cell **144**, 646–674 (2011)
26. E.L. Haseltine, J.B. Rawlings, Approximate simulation of coupled fast and slow reactions for stochastic chemical kinetics. J. Chem. Phys. **117**, 6959–6969 (2002)

27. B. Hat, K. Puszynski, T. Lipniacki, Exploring mechanisms of oscillations in p53 and nuclear factor-B systems. IET Syst. Biol. **3**, 342–355 (2009)
28. A.V. Hill, The possible effect of the aggregation of the molecules of hemoglobin on its dissociation curves. Proc. Physiol. Soc. **1**, 4–7 (1910)
29. M.W. Hirsch, S. Smale, R.L. Devaney, *Differential Equations, Dynamical Systems, and an Introduction to Chaos*, 3rd edn. (Academic, Amsterdam, 2012)
30. R. Hummel, D.J. Hussey, J. Haier, MicroRNAs: predictors and modifiers of chemo- and radiotherapy in different tumour types. Eur. J. Cancer **46**, 298–311 (2010)
31. G. Impicciatore, S. Sancilio, S. Miscia, R. Di Pietro, Nutlins and ionizing radiation in cancer therapy. Curr. Pharm. Des. **16**, 1427–1442 (2010)
32. K. Jonak, K. Jedrasiak, A. Polanski, K. Puszynski, Application of image processing algorithms in proteomics: automatic analysis of 2-D gel electrophoretic images from western blot assay, in *Computer Vision and Graphics*. Lecture Notes in Computer Science, vol. 7594 (Springer, Berlin, 2012), pp. 433–440
33. B. Kim, Q. Tang, P.S. Biswas, J. Xu, R.M. Schiffelers, F.Y. Xie, A.M. Ansari, P.V. Scaria, M.C. Woodle, P. Lu, B.T. Rouse, Inhibition of ocular angiogenesis by siRNA targeting vascular endothelial growth factor pathway genes: therapeutic strategy for herpetic stromal keratitis. Am. J. Pathol. **165**, 2177–2185 (2004)
34. M. Kracikova, G. Akiri, A. George, R. Sachidanandam, S.A. Aaronson, A threshold mechanism mediates p53 cell fate decision between growth arrest and apoptosis. Cell Death Differ. **20**, 576–588 (2013)
35. B. Krawczyk, K. Rudnicka, K. Fabianowska-Majewska, The effects of nucleoside analogues on promoter methylation of selected tumor suppressor genes in MCF-7 and MDA-MB-231 breast cancer cell lines. Nucleosides Nucleotides Nucleic Acids **26**, 1043–1046 (2007)
36. T. Lipniacki, K. Puszynski, P. Paszek, A.R. Brasier, M. Kimmel, Single TNFalpha trimers mediating NF-kappaB activation: stochastic robustness of NF-kappaB signaling. BMC Bioinf. **8**, 376 (2007). doi:10.1186/1471-2105-8-376
37. P.I. Makinen, J.K. Koponen, A.M. Karkkainen, T.M. Malm, K.H. Pulkkinen, J. Koistinaho, M.P. Turunen, S. Yla-Herttuala, Stable RNA interference: comparison of U6 and H1 promoters in endothelial cells and in mouse brain. J. Gene Med. **8**, 433–441 (2006)
38. L. Michaelis, M. Menten, Die kinetik der invertinwirkung. Biochemistry **49**, 333–369 (1913)
39. U.M. Moll, O. Petrenko, The MDM2-p53 interaction. Mol. Cancer Res. **1**, 1001–1008 (2003)
40. J. Momand, D. Jung, S. Wilczynski, J. Niland, The MDM2 gene amplification database. Nucleic Acids Res. **26**, 3453–3459 (1998)
41. J.D. Murray, *Mathematical Biology* (Springer, New York, 2002)
42. J.D. Oliner, K.W. Kinzler, P.S. Meltzer, D.L. George, B. Vogelstein, Amplification of a gene encoding a p53-associated protein in human sarcomas. Nature **358**, 80–83 (1992)
43. K. Puszynski, B. Hat, T. Lipniacki, Oscillations and bistability in the stochastic model of p53 regulation. J Theor. Biol. **254**, 452–465 (2008)
44. K. Puszynski, R. Bertolusso, T. Lipniacki, Crosstalk between p53 and nuclear factor-B systems: pro- and anti-apoptotic functions of NF-B. IET Syst. Biol. **3**, 356–367 (2009)
45. K. Puszynski, R. Jaksik, A. Swierniak, Regulation of p53 by siRNA in radiation treated cells: simulation studies. Int. J. Appl. Math. Comput. Sci. **22**, 1011–1018 (2012)
46. K. Puszynski, A. Gandolfi, A. d'Onofrio, The pharmacodynamics of the p53-Mdm2 targeting drug Nutlin: the role of gene-switching noise. PLOS Comput. Biol **10**(12), e1003991 (2014)
47. R.M. Raab, G. Stephanopoulos, Dynamics of gene silencing by RNA interference. Biotechnol. Bioeng. **88**, 121–132 (2004)
48. S. Shangary, S. Wang, Small-molecule inhibitors of the MDM2-p53 protein-protein interaction to reactivate p53 function: a novel approach for cancer therapy. Ann. Rev. Pharmacol. Toxicol. **49**, 223–241 (2009)
49. A. Sharei, J. Zoldan, A. Adamo, W.Y. Sim, N. Cho, E. Jackson, S. Mao, S. Schneider, M.J. Han, A. Lytton-Jean, P.A. Basto, S. Jhunjhunwala, J. Lee, D.A. Heller, J.W. Kang, G.C. Hartoularos, K.S. Kim, D.G. Anderson, R. Langer, K.F. Jensen, A vector-free microfluidic platform for intracellular delivery. Proc. Natl. Acad. Sci. U.S.A. **110**(6), 2082–2087 (2013)

50. M.S. Shim, Y.J. Kwon, Efficient and targeted delivery of siRNA in vivo. Fed. Eur. Biochem. Soc. J. **277**, 4814–4827 (2010)

51. H.S. Soifer, J.J. Rossi, P. Saetrom, MicroRNAs in disease and potential therapeutic applications. Mol. Ther. **15**, 2070–2079 (2007)

52. C. Tovar, J. Rosinski, Z. Filipovic, B. Higgins, K. Kolinsky, H. Hilton, X. Zhao, B.T. Vu, W. Qing, K. Packman, O. Myklebost, D.C. Heimbrook, L.T. Vassilev, Small-molecule MDM2 antagonists reveal aberrant p53 signaling in cancer: implications for therapy. Proc. Natl. Acad. Sci. U.S.A. **103**, 1888–1893 (2006)

53. L.T. Vassilev, B.T. Vu, B. Graves, D. Carvajal, F. Podlaski, Z. Filipovic, N. Kong, U. Kammlott, C. Lukacs, C. Klein, N. Fotouhi, E.A. Liu, In vivo activation of the p53 pathway by small-molecule antagonists of MDM2. Science **303**, 844–848 (2004)

54. B. Vogelstein, D. Lane, A.J. Levine, Surfing the p53 network. Nature **408**, 307–310 (2000)

55. P. Waage, C.M. Guldberg, Videnskabs-selskabet i christiana. Forhandlinger **35** (1864)

Chapter 6
Model Identification and Parameter Estimation

Abstract Analysis of the models presented in the preceding chapters may be focused on drawing conclusions of either qualitative or quantitative nature. In the first case, parameter values are not needed, as the goal is to determine, for example, stability properties or the form of the optimal control. Such conclusions subsequently provide the basis for quantitative analysis that concerns a particular cancer type and attempts to determine the outcome of a therapy, or, in more advanced studies, the optimal therapy protocol. The latter is of more value from the clinical point of view. However, applicability of modeling results depends on the ability to estimate correct parameter values for the models under consideration. In subsequent sections, estimation of the model parameters is discussed in the context of experimental and numerical procedures, relevant for the models described in previous chapters.

6.1 General Remarks

Computational models in systems biology, including those described in the preceding chapters, involve many unknown parameters. Their estimation is a challenging task whose completion is necessary for the models to be applied to support experimental or clinical research.

In general, parameter estimation requires solving a particular optimization problem, whose formulation depends on the model and available experimental data. Perhaps the most general way to describe such problem involves the Bayesian approach. The optimization goal, associated with parameter estimation, is then defined by finding estimators $\hat{\theta}$ of model parameter vector θ that maximize one of the following:

- probability of obtaining observations D, given the model M and its parameters θ (maximum likelihood criterion);
- posterior distribution of estimates $\hat{\theta}$, given observations D, the model M, and a prior distribution of $\hat{\theta}$;
- probability of a predicted model output, not observed experimentally, given experimental data and a set of feasible models with their parameter distributions (maximum averaged predictive likelihood criterion [28]).

© Springer International Publishing Switzerland 2016
A. Świerniak et al., *System Engineering Approach to Planning Anticancer Therapies*, DOI 10.1007/978-3-319-28095-0_6

The first case seems to be most natural. The parameter vector is given by

$$\hat{\theta} = \arg \max_{\theta} P(D|\theta, M). \tag{6.1}$$

Unfortunately, available data is usually scarce, since only few variables in high dimensional models are measured in few time points. This may lead to poor parameter estimates. To avoid this, prior knowledge about parameter distribution is required and maximum posterior probability estimate of parameter values seems to be more appropriate:

$$\hat{\theta} = \arg \max_{\theta} P(\theta|D, M). \tag{6.2}$$

From Bayes' theorem it follows that

$$P(\theta|D, M) = \frac{P(D|\theta, M)P(\theta|M)}{P(D|M)}. \tag{6.3}$$

Since the data prior $P(D|M)$ does not depend on parameter values, the solution is sought as

$$\hat{\theta} = \arg \max_{\theta} P(D|\theta, M)P(\theta|M), \tag{6.4}$$

where $P(\theta|M)$ is the probability density of the model parameters and represents prior knowledge.

If the model structure is not determined, the estimates may be found for each feasible model M_k. Then, in order to conclude that one model is better supported by experimental data than another, the ratio of marginal likelihoods $\frac{p(D|M_i)}{p(D|M_j)}$ needs to be calculated, where

$$p(D|M_k) = \int_{\theta_k} p(D|M_k, \theta_k)p(\theta_k|M_k)d\theta_k. \tag{6.5}$$

However, calculation of (6.5) is not a simple task. Various methods to accomplish this task have been proposed [28, 86].

When the model is fixed and the measurement uncertainty is described by a Gaussian noise, the maximization of the likelihood is equivalent to minimization of the mean square error between predicted and measured output. When such optimization goal is stated, we obtain [13]

$$\hat{\theta} = \arg \max_{\theta} \sum_{\alpha} \sum_{i_\alpha} \left(\frac{y_\alpha(t_{\alpha i}, \theta) - d_{\alpha i}}{\sigma_{\alpha i}} \right)^2, \tag{6.6}$$

where α is an index of the measured species, i_α is the number of measurements of species α, y is the simulation result, θ is a vector of parameter values, $d_{\alpha i}$ is the measured value at time $t_{\alpha i}$ for species α, and $\sigma_{\alpha i}$ is the measurement error. After the model has been successfully fitted to the experimental data, the resulting parameter estimate $\hat{\theta}$, frequently characterized by large variance(s) and covariances [13]. Though the variances decrease with increasing sample size, the latter is usually small in biological experiments. The covariances can be effectively approximated by the inverse of the Fisher information matrix (FIM) [29], providing information about which parameters contribute most to the variability of system behavior.

It appears that dynamics of complex biological systems often is determined by only a few *stiff* combinations of the parameters and changes of other *sloppy* parameter combinations, even by orders of magnitude do not influence general system behavior [29]. This property is particularly relevant when the modeling is focused on intracellular processes (see also Sect. 6.4). Then, optimization algorithms which adapt to diverging step sizes along different parameter combinations are needed (such as Levenberg–Marquardt and Nelder–Mead methods) [13]. After finding the best fitting set of parameters, prediction uncertainties are calculated by accounting for model behavior over the region of parameter space that is consistent with available data, e.g. applying linearized covariance analysis (LCA) or Monte Carlo analysis (MCA) [30].

Ideally, one should aim at designing experiments to directly measure parameter values. However, despite advances in experimental techniques, this is not always possible. Moreover, mathematical analysis of sloppiness using Vandermonde matrices [83] and the comparisons in [29] indicate that models with "sloppy" parameters constitute rule not exception. Therefore, "predictions are generally much more efficiently constrained by collectively fitting model parameters than by directly measuring them" [29].

6.2 Compartmental Population Models

Most parameters in compartmental models describe flux rates from one compartment to another. Their biological interpretation, however, depends on the model type as well as the experimental techniques used to estimate the parameters. Taking into account the models presented in Chaps. 2 and 3, the parameters under consideration correspond either to the lengths of particular cell cycle phases, or fractions of cells arrested and/or killed by drugs, or probabilities associated with the events leading to the change in the gene copy number or missense point mutations. Estimation of each of these parameters requires different experimental and computational methods.

6.2.1   Length of Cell Cycle Phases

For estimation of parameters related to cell cycle and fraction of cells that are in a particular phase of the cell cycle or are apoptotic or necrotic, flow cytometry has been the main experimental technique since ca. 1970s. As immunocytochemical detection of cyclins D, E1, A2, and B1 became possible, in addition to DNA content measurement, six to eight cell cycle compartments differing in their degree of progression through the cycle, G_0/G_1, S, G_2, Prophase, Prometaphase, Metaphase, Anaphase, Telophase, and cytokinesis, can now be distinguished [18]. On the other hand, administration of thymidine analogues to tumor-bearing animals or cancer patients has enabled taking measurements in vivo [14].

Recently, dynamic microscopy imaging has been developed to support analysis of living cells. In particular, a single-cell tracking approach has been developed that enables automatic detection of cell cycle phases using fluorescent ubiquitylation-based cell-cycle indicator (FUCCI) probes [59, 69]. Special software is used to automatically track the cycle progression of individual cells and their division [22]. Other reporter systems are also being developed [17]. These tools not only facilitate estimation of the length of particular cell cycle phases but also provide information about their variability in what might appear a homogeneous cell population. Moreover, when combined with other reporter systems, they allow to investigate correlations of cell cycle responses to external stimuli. In a recent paper [21], we estimated parameters in a stochastic model of cell cycle control, based on single-cell experiments. This is a follow-up to earlier estimations carried out using the bifurcating autoregression approach [67].

6.2.2   Rates of Missense Mutational Events

Numerous studies have shown that missense mutations play an important role in carcinogenesis. Their overall effect has been shown to be destabilizing, mostly affecting the electrostatic component of binding energy [48]. These discoveries are enhanced by bioinformatics analysis applied in cross-species research to estimate the rates of mutational events. Progress in this area is fueled mostly by large-scale sequencing techniques that have been developed in recent years, allowing identifying mutations present in different cancer types. They have been supported by theoretical studies aimed at identification of probable mutation sites [5, 19, 23].

While large-scale sequencing projects combined with phylogeny reconstruction provide information about the mutation rates in general, not all mutations contribute to carcinogenesis. When the research is focused on development of models of time-dependent evolution of cancer populations and their responses to treatment, it is only a subset of mutations that is of interest. Therefore, numerous methods have been developed to predict the impact of missense mutations on protein structure and function, such as, among others, SIFT [47], Align-GVGD [41, 71], PolyPhen-2

[1, 55], and FIS [58]. It has been shown, however, that these algorithms make different predictions, depending on algorithm parameters and sequence alignment that precedes their initiation [32]. Therefore, caution is needed when using their output as the source of parameter values for dynamical models.

Two parameters are usually calculated when considering missense mutations. The first of them is the highly variable rate of nonsynonymous substitutions per nonsynonymous substitution site (K_A). This variability is attributed to different patterns of selection acting at different sites. The other is the much less variable rate of synonymous substitutions (considered neutral) per synonymous substitution site (K_S), related to the average mutation rate for the genome as a whole. The K_A/K_S ratio is a measure of accepted substitutions normalized for opportunity and can indicate whether or not selection is occurring and help to determine the degree and mode of selection ([73] and the references therein). Multiple algorithms have been developed to estimate these rates, for fast-evolving, intermediately evolving, and slow-evolving genes [81]. They have been subsequently used to model mutation acquisition in a normal tissue and to investigate how deregulation of the mechanisms preserving tissue homeostasis contributes to cancer [27]. Interesting patterns of mutations have been described in the yeast model of breast cancer mutations [68, 85].

6.2.3 Gene Copy Number Variation

Acquisition and/or loss of additional gene copies may significantly alter cell behavior and its responses to environmental factors. When this phenomenon is included either in proliferation of cancer cells or in response to treatment, one should estimate the number of gene copies and its distribution in a cell population on one hand, and probabilities of acquiring and losing gene copies on the other. For the former task a number of techniques can be employed, among others comparative genomic hybridization, fluorescent in situ hybridization, digital karyotyping, quantitative microsatellite analysis, and BAC end sequencing [2]. They clearly prove that the number of additional gene copies may be large and highly variable, therefore constraining models to include only several of them is not enough.

Experimental works devoted to analysis of gene amplification-related drug resistance date back at least to the early 1980s (see, e.g., [10, 34, 44, 45, 74, 80]). They provide estimates for probabilities of a relatively rare primary event, i.e. the establishment of the founder cell of the resistant clone containing at least one unstable copy of the target gene as well as subsequent amplification and deamplification events. As for the probability of the primary mutational event, estimates ranged from 10^{-8} to 10^{-6} (the probability of this event, per cell division, corresponds to the ratio α/λ in the model described by (2.52)). Subsequent amplification and deamplification events occur at higher rates, as a result of instability of the amplified gene. The probabilities of these events, per cell division, correspond to the ratios b/λ

(amplification) and d/λ (deamplification) in (2.52) and are of the order of 0.02 and 0.10, respectively [36].

Currently, next-generation sequencing, SNP arrays, and array comparative genomic hybridization platforms combined with statistical analysis, using among other variational Bayesian mixture models, are used to estimate the parameters mentioned above, separately for each type of cancer being investigated ([42] and the references therein).

6.3   Structured Models, Models Including Angiogenesis and Spatial Models

These models are the ones for which estimating parameters seems to be the most challenging task. The main reason is that, possibly except for leukemias, to be relevant, parameter estimation in these models must be done in vivo. Therefore, available data concern either tumor development in mice, in which case it is possible to trace time course of some variables, or magnetic resonance (MR) imaging of human cancers that shows the structure of blood vessels or shape of a solid tumor. For example, a simple Gompertzian model that describes in vivo tumor growth and its sensitivity to treatment with antiangiogenic drugs was analyzed in [24]. Two parameters were estimated there, one relating to the initial mitosis rate and another one relating to the deceleration of growth caused by antiangiogenic factors. According to our knowledge, there are no published results on experimental parameter estimation for models of antiangiogenic therapy.

Only recently new experimental techniques have been developed to support investigation of vasculature in vitro, which could provide data necessary to estimate parameters for the models describing the dynamics of vasculature growth [75, 78].

On the other hand, MR imaging provides information that can be used to verify simulation results of the spatial models showing growth of vasculature [76], allowing to assess such characteristics as number of branches per volume unit and vessel diameter in different areas of the tumor.

6.4   Models of Signaling Pathways and Intracellular Processes

The constituent chemical species and the pathway structure are frequently not fully characterized, due to lack of biological knowledge. Even if the influence network is fully known, it may be so complex that simplification is needed because of the computational power limits. Therefore, it is necessary to decide the best model structure. As argued by Girolami: "Bayesian inferential methodology provides a coherent framework with which to characterize and propagate uncertainty in such mechanistic models" [28]. His paper provides an outline of Bayesian methodology for system models represented as differential equations. In another paper by Girolami's group [86] a similar methodology is applied to ranking of alternative hypothetical topologies of a cell signaling pathway. In a more recent work [15], Chkrebtii et al. develop a fully Bayesian inferential framework to quantify uncertainty in models defined by systems of analytically intractable differential equations. The approach is successfully applied to ordinary and partial differential equation models and to an example characterizing parameter and state uncertainty in a biochemical signaling pathway which incorporates a nonlinear delay-feedback mechanism.

Besides the structure identification problem it is also very difficult to experimentally obtain values of the chemical reaction constants which are the parameters of the mathematical model. It is possible to find the parameter values by fitting the model to data on mRNA or protein levels, such as discussed in Sect. 5.1.2. Unfortunately, estimation of parameters by means of data fitting is a problematic task when the number of parameters is large as in models of large biochemical network (see, e.g., [52]), and even more problematic if the model is fully stochastic or hybrid such as described in Sect. D.2.3.

One of the common problems with parameter estimation is model sloppiness characteristic for the models in systems biology which usually involve many free parameters [29].

The second common problem is structural or "practical" non-identifiability of the model parameters which is the main obstacle to using automatic fitting algorithms such as least-squares minimization algorithms or simulated annealing. We must also recall that the existing methods of verifying identifiability even for deterministic models become rapidly unfeasible as the size of the ODE model increases [49]. If the model is not identifiable, the minimum of the performance index is surrounded by a large flat region or multiple local minima of comparable "depth" [57]. In such cases, from the identifiability point of view, time-consuming algorithms such as those based on least squares minimization do not have advantage over Monte Carlo or even trial-and-error method. Moreover, even if a sophisticated global search algorithm could find the true minimum, but usually the user has to select a box in the parameter space in which the search will take place, the "uniqueness" of that parameter set is questionable on the grounds of biological meaning. This uncertainty should be taken into account when the model is used to prevent unproven conclusions about the biological system, or in making overly optimistic predictions without understanding the uncertainty of the model.

Estimation of parameters associated with kinetics of intracellular processes requires algorithms and approaches tailored to specific models. While there exist a variety of experimental procedures that can be employed to trace the amount of molecules of a specific type in cell cultures, all of them have drawbacks. These procedures can be divided into two general categories:

- population-level experiments, employing blotting techniques, microarrays, PCR, etc.—amounts of a very large set of molecule types can be measured; however,

the results are averaged over population and do not provide valuable information if the response dynamics varies significantly from cell to cell (see further on);
- live cell imaging; it facilitates observation of living cells, heterogenous responses—however, usually only up to two variables are measured in a given cell population and the variety of molecules that can be traced this way is much smaller.

6.4.1 Degradation and Transcription Rates

Among parameters associated with processes described by models, only the degradation rates seem to be those that can be relatively easily estimated. Experiments, in which production of a certain transcript or protein is blocked, are now routine (see, e.g., [70]). Assuming that the degradation is a first-order process, degradation rate can be determined from a series of measurements (see Fig. 6.1).

Estimation of transcription rates is more complex, but recent advances in live cell imaging have significantly increased its accuracy. Live cell fluorescent reporter-based techniques reveal the dynamics of gene expression under the control of different regulatory promoters, in individual cells and over periods of several days. Experiments with destabilized reporters facilitate observations of when genes are turned on, how long their expression lasts and show possible periodic or random repetitions [22, 31, 84].

6.4.2 Population vs Single Cell Experiments

The concept of deterministic modeling and fitting model parameters based on Western blots, EMSA blots, real-time PCR, and other data is based on the hypothesis of homogeneity of the cell population. However, it is known that even if all cells are cloned from the same ancestor, they are not homogenous in the sense of dynamics of

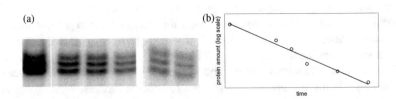

Fig. 6.1 (a) An example of experimental results showing decreasing concentration of a given molecule type; (b) quantified data (*circles*) and simulation results obtained for a model whose parameters were estimated using the linear regression method (in semi-logarithmic coordinate system the exponential curve of degradation becomes a straight line). Similar calculations can be performed also with data from live cell imaging

intracellular processes. The heterogeneity can be attributed to stochastic distribution of initial conditions (e.g., the number of molecules of a given type at time 0) or the differences in kinetic parameters of processes analyzed. As a result, different dynamics of intracellular processes may be observed. This may lead to deceptive artifacts, considering that experimental procedures consist in growing cell cultures on different plates, each of which provides a measurement for one time point (see Fig. 6.2b). Due to differences in initial conditions and kinetic constants for different cells (even of the same type) such procedures lead to quantitatively (but not qualitatively) different responses in cell subpopulations. System response can be distorted falsely implying oscillatory behavior.

If there are large oscillations in single cells, data gathered in cell population experiments do not reflect real dynamics and therefore cannot be used for parameter fitting. When cells are exposed to external excitation, the initial response can be synchronized. However, in the course of the experiment, cells lose their initial synchrony, and the oscillations characteristic for the dynamics of a given pathway are lost through the averaging experiment (Fig. 6.2a). Once again, as in the preceding subsection, the problem lies in heterogeneity of the cell population. As a consequence, single cell experiments may be required, in which individual cells in a population are observed with a microscope, and proteins or other molecules of interest are tagged by fluorescent particles (see, e.g., [87]).

There are two additional drawbacks to the population methods. First, due to substantial costs of experiments, usually the variables are measured only at several time points. Their choice may significantly alter conclusions drawn from the measurements, sometimes totally distorting the kinetics of the processes involved. Second, dynamics of fast processes cannot be captured at all using these techniques, as sample preparation takes at least several minutes, during which some of the processes are still running.

Despite all these drawbacks, several new methods have been developed recently to improve the quality of experimental results obtained at cell population level. One of them is stable isotope dilution (SID)-selected reaction monitoring (SRM)-mass spectrometry (MS) assays for quantification of low abundance signaling proteins

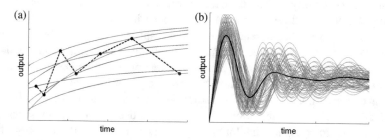

Fig. 6.2 Dynamical artifacts in experimental data, when procedure does not trace the behavior of individual cells: (**a**) misleading oscillations observed when the replicate cell plates have been used in the order shown by numbers in the plot; (**b**) oscillations in individual cells are canceled on the population level

[90] which could not have been observed otherwise, including live microscopy. This approach has been applied, e.g., to build and analyze a model of innate immune response to dsRNA stimulation [7].

Live microscopy imaging presents an alternative to the population averaging techniques. Various methods have been developed to allow observation of molecules of interest in individual, living cells. Fluorescence Recovery after Photobleaching (FRAP) is one of the most common. It consists in irreversibly photobleaching fluorescent molecules by high intensity illumination with a focused laser beam. Subsequently, this fluorescence is recovered with a particular velocity, recorded at low laser power. Temporal fluorescence changes illustrate changes in molecular content in the area of interest [33]. Other methods based on the same general idea have been developed in recent years, including, among others, FLIP, FLAP, FLIM [33], and s-FDAP [82]. A review of these and other methods can be found in [20].

6.4.3   Relative vs Absolute Measurements

The models of signaling pathways most often use molar concentration of molecules, or the average number of molecules of a given type, as state variables. However, the experiments frequently result in only relative measurements. There exist virtually no databases containing information about the actual content of cells in terms of typical concentrations of known proteins, complexes, or other molecules. Though it is in principle possible to perform experiments that provide such data, the cost involved is too high and as a result in most cases data available allows to compare only the levels of a given molecule to the initial level of molecules of the same type. Therefore, to allow comparison of experimental data coming from different sources and simulation results, normalization of the results, both experimental and numerical, is necessary.

However, normalization restricts research to qualitative analysis only, as systems exhibiting the same type of dynamics, but differing with respect to the values of parameters, after normalization may be indistinguishable (Fig. 6.3). Therefore, one should also check original results before normalization to avoid mistakes.

Single-cell methods such as FRAP, mentioned in the preceding section, while facilitating capturing heterogeneity in cellular responses, also provide only relative measurements. It is only recently that full quantification has been applied in experiments that measure protein and/or mRNA content of cells. The fluorescent in-situ hybridization (FISH) method has been in place for some time to provide accurate integer counts of mRNA molecules in individual cells. On the other hand, combination of observation of luminescence decay in cell populations after translation or transcription inhibition with serial dilutions of the reporter allows to calibrate the experimental system [70]. However, relatively few studies that couple modeling with experimental data of these types have been published (e.g., [11, 46, 54]). Nevertheless, this seems to be a large step forward in attempts to measure accurately absolute levels of molecules involved in intracellular control

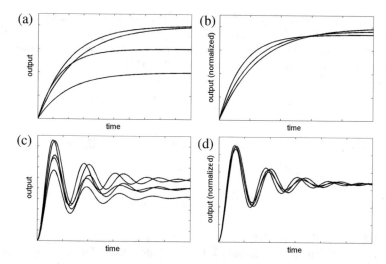

Fig. 6.3 Effects of normalization on results comparison: when normalized to the area under the plot, most of the variation disappears: (**a**) original and (**b**) normalized oscillatory response, and (**c**) original and (**d**) normalized aperiodic response

processes. If these techniques are included in standard experimental procedures, it will be possible to estimate reactions kinetic parameters with much better accuracy and confidence.

6.4.4   Algorithms Used to Estimate Kinetic Parameters

There are several approaches that can be employed to estimate parameter values once the experimental data are quantified. The most intuitive one, in continued use because of the ill-posedness of the mathematical optimization problem arising in parameter identification for large systems with many unknown parameters, is manual fitting of the simulation and experimental results, supported by the expert knowledge of researchers evaluating the fit quality (see, e.g., [7, 64]). Other methods are based on solving the problem given by expression (6.6). In [89] the use of Hartley's modulating function was proposed, but the approach is limited to simple nonlinear models under the assumption that measurements are dense in time. Another possible method involves application of generalized backpropagation through time (GBPTT) algorithm for training continuous-time neural networks with discrete-time measurements and adjoint sensitivity analysis. The adjoint system provides the gradient of a quadratic performance index and is used for fitting of parameters of the model [25, 26]. If all state variables describing the dynamics of a biochemical processes network are observed, Bayesian inference methods may be applied, providing not only parameter estimates but also quantifying their statistical relevance. For example, they allow to obtain the predictive probability distribution of the unobserved random variable y_*, $p(y_*|u_*, D)$ and provide the probability distribution of the values which y_* may take, given, or conditioned upon, the known value of input u_* and the previously observed data D [28]:

$$p(y_*|u_*, D) = \sum_k p(y_*|u_*, M_k, D)P(M_k|D), \tag{6.7}$$

where $P(M_k|D)$ defines the posterior probability of model M_k given the observed data

$$p(M_k|D) = \frac{p(D|M_k)\pi(M_k)}{\sum_i p(D|M_k)\pi(M_k)} \tag{6.8}$$

with $\pi(M_k)$ and $p(D|M_k)$ (defined by (6.5)) denoting the prior probability distribution over models and the integrated likelihood of the observed data given the model, respectively. The model specific predictive likelihood $p(y_*|u_*, M_k, D)$ is given by

$$p(y_*|u_*, M_k, D) = \int_{\theta_k} p(y_*|u_*, \theta_k, M_k)p(\theta_k|D, M_k)d\theta_k, \tag{6.9}$$

where

$$p(\theta_k|D, M_k) = \frac{p(D|\theta_k, M_k)p(\theta_k|M_k)}{p(D|M_k)}. \tag{6.10}$$

6.5   Pharmacokinetics Parameters

Estimation of parameters for pharmacokinetic models adds a new twist to the problem. In general, they are estimated in clinical trials, in which drug concentration can be measured in plasma, subcutaneously, or, in some cases, using MR or computed tomography (CT) imaging. From the clinician's point of view, "pharmacokinetic parameter" is not the parameter in a mathematical model, but an interpretable quantity, e.g. [3, 4, 8]:

- minimum inhibitory concentration (MIC);
- maximum and minimum concentrations during the dosing interval;
- clearance rate;
- volume of distribution at steady state—the apparent volume in which a drug is distributed following intravenous injection with equilibration between plasma and the surrounding tissues;
- area under the curve (AUC) (after a single dose or in steady state);

- mean residence time (MRT)—the amount of time that a medication spends in the body;
- terminal half-life time required for the concentration to fall by 50 % after reaching equilibrium.

The above clinical parameters are subsequently used to estimate the input value for the models describing drug actions which is the problem of parameter estimation in dynamical models. Nevertheless, numerous algorithms have been developed with this goal in mind, in particular aimed at predicting the ratios of tissue and blood, or tissue and plasma concentrations, for total or unbound drug at macro (i.e., whole tissue) and micro (i.e., cells and fluids) levels [51].

Heterogeneity of pharmacokinetics in patient population is considered by the so-called population PK models. Usually pharmacokinetics, describing the relationship between the drug dose and its concentration in the plasma in a single patient (see Sect. 2.3), is described by very simple two-compartment models. Then, with support from clinical observations from a group of patients, nonlinear mixed effects modeling with first-order conditional estimation with interaction method can be applied [39]. Multi-compartmental models are also employed, with compartments corresponding to different organs or tissues. Monte Carlo methods and multilevel population models are applied to gain an insight into the determinants of population heterogeneity and differential susceptibility of patients in so-called physiologically based pharmacokinetic (PBPK) modeling [9]. Bayesian reasonig is then used to the estimate effects of different individual factors on the pharmacokinetics of a given drug [50, 60]. Bayesian methods are used not only for parameter estimation but at the same time they can be used to support experiment planning [77].

6.6 Sensitivity of Regulatory Network Models

Another tool, besides the bifurcation analysis, used to determine how the change of parameters influences system behavior is the sensitivity analysis. It helps to identify parameters which have the greatest impact on the system output both in steady and transient states. Moreover the sensitivity analysis provides information about the systems robustness, in the sense of range of parameter values for which qualitative system responses remain unchanged. This property should characterize most of the signaling pathways as cells respond similarly to external stimuli despite the inherent heterogeneity in a cell population.

Interpretation of the results of application of sensitivity methods to signaling pathway models goes beyond standard sensitivity/robustness conclusions. Among others, it also provides means to simplify high dimensional models that arise in systems biology [35] and can be used to indicate prospective molecular targets for the drugs against diseases associated with particular signaling pathways [40].

Sensitivity analysis methods fall into two categories: local and global. Local sensitivity analysis is usually defined as analysis of the influence of a small, local deviation of a single parameter on the system outputs. Global sensitivity analysis deals with the influence of multiple parameter changes, over a wide range.

Description of all the known sensitivity methods is beyond the scope of this work. A good overview of their applicability can be found in reference [79]. We briefly describe the local sensitivity analysis based on sensitivity matrix and the global sensitivity analysis based on Sobol indices.

6.6.1   Local Sensitivity Analysis

Let us consider the system output being identical with its state and assume a model described by the state equation:

$$\frac{dX}{dt} = f(X, u, p) \tag{6.11}$$

whose solution is

$$X(t) = X(p_n, t), \tag{6.12}$$

where p_n denotes the nominal (i.e., chosen for calculations and usually assumed to represent the mean in a cell population) parameter vector. The first-order sensitivity coefficients s_{ij}, describing the influence of the i-th parameter on the j-th state variable are defined as [16, 61]:

$$s_{ij} = \frac{\partial x_i}{\partial p_j}\Big|_{p_n} \tag{6.13}$$

and the absolute sensitivity matrix as

$$S = \frac{\partial X}{\partial p}\Big|_{p_n} = \begin{bmatrix} s_{11} & s_{12} & \cdots & s_{1m} \\ s_{21} & s_{22} & \cdots & s_{2n} \\ \vdots & \vdots & \vdots & \ddots \\ s_{n1} & s_{n2} & \cdots & s_{nm} \end{bmatrix}. \tag{6.14}$$

More precisely, s_{ij} are sensitivity functions, as they change in time.

Usually the analytical solution of Eq. (6.12) is not available, therefore sensitivity coefficients must be calculated in some other way. One of the approaches is the direct differential method [37]. Differentiating (6.11) with respect to p_j, we obtain

$$\frac{d}{dt}\frac{\partial X}{\partial p_j} = \frac{\partial f}{\partial X}\frac{\partial X}{\partial p_j} + \frac{\partial f}{\partial p_j} = J \cdot S_j + F_j, \tag{6.15}$$

where

$$
J = \frac{\partial f}{\partial X} =
\begin{bmatrix}
\frac{\partial f_1}{\partial x_1} & \frac{\partial f_1}{\partial x_2} & \cdots & \frac{\partial f_1}{\partial x_n} \\
\frac{\partial f_2}{\partial x_1} & \frac{\partial f_2}{\partial x_2} & \cdots & \frac{\partial f_2}{\partial x_n} \\
\vdots & \vdots & \ddots & \vdots \\
\frac{\partial f_n}{\partial x_1} & \frac{\partial f_n}{\partial x_2} & \cdots & \frac{\partial f_n}{\partial x_n}
\end{bmatrix}
\tag{6.16}
$$

is the Jacobi matrix,

$$
F_j = \frac{\partial f}{\partial p_j} =
\begin{bmatrix}
\frac{\partial f_1}{\partial p_j} \\
\frac{\partial f_2}{\partial p_j} \\
\vdots \\
\frac{\partial f_n}{\partial p_j}
\end{bmatrix}
\tag{6.17}
$$

is the parametric Jacobi matrix, and

$$
S_j = \frac{\partial X}{\partial p_j} =
\begin{bmatrix}
s_{1j} \\
s_{2j} \\
\vdots \\
s_{nj}
\end{bmatrix}
\tag{6.18}
$$

is the column sensitivity vector with respect to the j-th parameter.

Finally, Eqs. (6.11) and (6.15) are combined and solved simultaneously to find sensitivity coefficients:

$$
\begin{cases}
\dot{X} = f(X, p, u, t) \\
\dot{S}_j = J \cdot S_j + F_j
\end{cases}
\tag{6.19}
$$

An important problem which should be addressed here is how to correctly choose the initial conditions for the sensitivity functions. The common approach is to set them equal to 0. However it is true only in cases when arbitrary initial conditions $x_i(0)$ are assumed. When the task is to analyze the response of a system whose initial state is its equilibrium, corresponding to another input value, initial conditions depend on model parameters and the initial conditions for sensitivity functions are calculated using the formula:

$$
S_j(0) = \frac{\partial x(0)}{\partial p_j}.
\tag{6.20}
$$

Because in many systems different parameters as well as model state variables may take values that are distributed over several orders of magnitude, using relative sensitivities instead of the absolute sensitivities may be more appropriate. The latter are defined as

$$\bar{s}_{ij} = \frac{\partial x_i}{\partial p_j} \cdot \frac{p_j}{x_i} \tag{6.21}$$

Such normalization makes possible to compare relative influence of any parameter change on system behavior, regardless of the scale of either the parameter or the state variable.

After calculation of the relative sensitivities, the analysis can be focused on either of the following goals [53]:

- checking which parameters are relevant for the steady state and which for transient dynamics—the whole time course of sensitivity coefficients is taken into account;
- creating the ranking of parameters that indicates which processes are most important for the signaling pathway; consequently, this provides valuable information for experimental research about possible molecular targets in the pathway under consideration;
- finding correlation among parameters, which is important if the experiments are designed to estimate model parameters.

6.6.2 Global Sensitivity Analysis

Although the local sensitivity is useful, it has significant drawbacks. It is usually based on the assumption that only one parameter is changed in time in a limited range. What we are interested in is development and analysis of signaling pathway models in real cells. These cells are not identical but may differ in size, cell cycle phase, enzyme concentrations, or may be exposed to different environmental conditions. This is reflected in the parameter values which may be changed simultaneously for multiple parameters in a wide range. Depending on the system structure, some of these changes may increase and some may decrease the effect compared to the change of a single parameter. Therefore, global sensitivity analysis should be applied. Most of such analyses rely on simulation of the model for a large number of parameter sets and subsequent transformations of the results. What is important, their applicability is not constrained to deterministic models; stochastic models can also be analyzed in this way. Because in this type of sensitivity analysis a wider space of parameters is considered, the key issue is appropriate sampling of the parameter space. Many works devoted to this subject can be found in the literature, dealing with theoretical aspects of Latin hypercube sampling or factorial sampling plans (see, e.g. [12, 43, 62]), and their application to signaling pathways analysis (e.g., [88, 91]).

Two possible distributions are used to randomly generate the signaling pathway model parameters:

- uniform distribution, defined on a wide range of biologically acceptable parameters, if the parameters are not known.
- normal or Gamma distribution, if the nominal value of the parameter has already been determined experimentally or known from literature;

Subsequently, one of the two families of methods may be applied:

- variance-based sensitivity methods, where

  - the Fano factor, i.e. ratio of variance of a model output to the average value is calculated, serving as a sensitivity index [56], or
  - variance of a model output is decomposed into partial variances contributed by changes in the individual parameters; the sensitivity indices are subsequently derived from the ratio of the partial variance to the total variance of model output. [35, 66].

- calculating local sensitivities for each simulation and subsequently averaging them over all simulations [6, 35];

Another example of variance-based approaches is the method of Sobol's indices [66].

Let us assume, as before, that the model is described by the state Eq. (6.11), whose solution is (6.12). Let us also arbitrarily divide the set of all M parameters into two subsets y and z such that

$$y = (p_{k_1}, \ldots, p_{k_m}),\qquad(6.22)$$

where $1 \leq m \leq M - 1$, $1 \leq k_1 \leq \cdots \leq k_m \leq M$, and z contains the remaining $M - m$ parameters. Then, two sensitivity indices for each subset y can be defined:

$$S_y = \frac{D_y}{D},$$
$$S_y^{\text{tot}} = \frac{D_y^{\text{tot}}}{D}\qquad(6.23)$$

where D is a total variance of the model response to feasible changes in parameter values and D_y is the variance of the model response when only parameters from the subset y change. D_y^{tot} is the variance in the case when at least one of the changed parameters belongs to y. The total variance D is obtained by summing all possible D_y.

Because in most cases the analytical determination of the D_y, D_y^{tot}, and D values is impossible, a numerical approach based on Monte Carlo simulations is required. The Monte Carlo-based algorithm developed by Sobol in [66], to determine values of the variances, requires we divide the set of the parameters into two subsets (y, z), where y is the subset of the parameters for which we want to calculate Sobol indices and z is the subset containing the remaining parameters. Next step is to sample two points from the parameter space from the uniform distribution over the interval $[0, 1]$ to receive $P = (y, z)$ and $P' = (y', z')$. The assumption about the uniform distribution and its range is needed for convergence of the method and requires appropriate rescaling of the parameters in the models analyzed. Then three simulations of the model are run, for parameter sets $P = (y, z)$, $P_1 = (y, z')$ and $P_2 = (y', z)$. If we assume that $x(P)$ is the model simulation result for parameter set P, for example a protein level at time T, then after N simulations we receive

$$\frac{1}{N} \sum_{i=1}^{N} x(P) \rightarrow x_0 \tag{6.24}$$

$$\frac{1}{N} \sum_{i=1}^{N} x^2(P) \rightarrow D + x_0^2 \tag{6.25}$$

$$\frac{1}{N} \sum_{i=1}^{N} (x(P) \cdot x(P_1)) \rightarrow D_y + x_0^2 \tag{6.26}$$

$$\frac{1}{N} \sum_{i=1}^{N} (x(P) \cdot x(P_2)) \rightarrow D_z + x_0^2 \tag{6.27}$$

which allows determination of D, D_y, and D_z when N is large enough (the symbol $\rightarrow$ denotes stochastic convergence). Then we calculate Sobol indices for a chosen subset y of the parameters set by using Eq. (6.23) and formula:

$$D_y^{\text{tot}} = D - D_z \tag{6.28}$$

6.6.3   Parameters Ranking

As mentioned before, one of the possible applications of the sensitivity analysis of the signaling pathway models, especially those related to apoptosis is to find the best possible target for the anticancer therapies if these therapies are to result in a given molecule's level change. To accomplish this inside the cell, we should influence reactions, such as mRNA degradation by using matching siRNA, mRNA translation by using miRNA, or complex creation by using interfering biomolecules. Of course, the best targets are those to which the model is most sensitive, because the same percentage change of their rate results in higher output signal change. Assuming that for this same percentage change, the same amount of the drug with the same toxicity is required, targeting reactions to which the model is most sensitive means smaller drug doses and smaller side effects.

As far as the local sensitivity is concerned, numerical solution of (6.19) with the normalization as in (6.21) if necessary gives us the sensitivities of the particular time moments. Ranking of the parameters is therefore based on cumulative indices. These can be calculated either for each state variable separately or for the whole system.

The importance of the j-th parameter for the i-th state variable can be measured as [53]

$$S_{ij}^* = \frac{1}{T} \int_0^T |s_{ij}(\tau)| d\tau, \qquad (6.29)$$

for continuous systems, where T denotes the time horizon of the simulation, or [88]

$$S_{ij}^* = \frac{1}{N} \sqrt{\sum_{k=1}^N |s_{ij}(k)|^2}, \qquad (6.30)$$

for discrete systems, where N denotes the number of integration steps in the simulation and the sum is calculated over the consecutive values of s_{ij} from simulation.

Similarly, the overall effect of the j-th parameter change on the whole system can be expressed as

$$S_j^{\text{tot}} = \sum_{i=1}^n S_{ij}^*, \qquad (6.31)$$

For the global sensitivity analysis based on the Sobol method, the Sobol indices provide the influence of the selected subset of parameters on the simulation results, as opposed to influence of a single parameter. Moreover, any single parameter can be included in more than one subset (subsets generally overlap) and the results for these subsets may differ. Therefore, to estimate the total influence of the specific parameter on the model response, we have to consider all the subsets in which this parameter can be included. Similarly, if we want to construct the ranking of all parameters, we have to calculate Sobol indices for all possible subsets y of the parameter set, except for two cases. The first is an empty y subset, in which case Sobol indices are always equal to 0 and the second is when all parameters are included in y and the set z is empty. In these cases, the Sobol indices are always equal to 1.

When applied to the analysis of time responses of a dynamical system, the Sobol indices become functions of time. As a result, a single value of the index does not constitute a measure of the influence of parameter changes on system response. Therefore, the average time values of the indices should be chosen as the partial scores L_k for all parameters that belong to a given subset y_k. Because the y_k subsets have various sizes, it is reasonable to consider the influence of the partial score for a given parameter depending on the size k of the subset y_k, for which the partial score was received. In this way, smaller subsets have higher influence on the final score.

According to this the final score for a single parameter r could be, for example, calculated from the formula:

$$J_r = \sum_{k=1}^{M-1} \frac{1}{k} \cdot L_k^{(r \in y)},$$ (6.32)

where M is the size of the parameters set and $r \in y$ means that above sum is calculated for all L_k received from y_k that contain the parameter r.

After calculation of all the J_r values, the ranking of the parameter values is built. A higher J_r means that the model response is more dependent on parameter r and this parameter receives higher position in the ranking.

It should be noted that the Sobol indices can also be used for model verification and reduction. If Sobol indices for a single parameter are equal to 0, then reactions associated with these parameters can be omitted without any consequences on models response. Similarly when Sobol indices for single parameter changes are equal to 1, then the response depends only on this parameter and the remaining parameters can be set to 0 and reactions associated with them can be omitted.

6.6.4   Numerical Examples

In order to illustrate the concepts introduced in this section, let us consider three examples, focusing on different aspects of sensitivity analysis. Let us start with a standard second order oscillatory system, described by the following state equations:

$$\begin{aligned} \dot{x_1} &= x_2 \\ \dot{x_2} &= -p_1 u(t) - 2\xi \omega_n x_2 - \omega_n^2 x_1 \end{aligned},$$ (6.33)

where $u(t)$ is the system input and parameters k, ξ, and ω_n represent system gain, damping factor and natural undamped frequency, respectively.

Assuming arbitrary nominal parameter values, it is possible to calculate sensitivity functions (6.13). These functions directly show how small parameter changes influence system response (Fig. 6.4).

However, in many biological models, both parameters and variables may differ in the orders of magnitude. Therefore, normalized sensitivity functions (6.21) are more useful, though they cannot be used in the same way to calculate system responses (Fig. 6.5).

Sensitivity functions may be used to calculate parameter rankings, as presented in Sect. 6.6.3. For example, if the system (6.33) is considered, the ranking shows that the gain k is the most important parameter, as illustrated in Fig. 6.6.

To illustrate simple, but important conclusions stemming from the application of sensitivity functions let us consider one of the models of NFκB pathways, described in [38]. Without going into details, it consists of 13 ODEs and 28 parameters. The

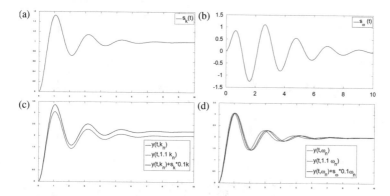

Fig. 6.4 (**a**),(**b**) Sensitivity functions for a second order oscillatory system calculated with respect to parameters k and ω_n, respectively, and comparison of system responses obtained for nominal parameters (*blue*), changed parameter (*red*) and calculated for changed parameters with sensitivity functions (*black*) with changed parameter (**c**) k and (**d**) ω_n

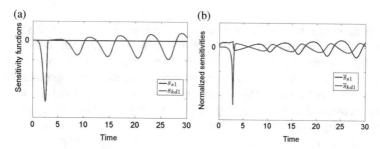

Fig. 6.5 An example of (**a**) sensitivity function (**b**) normalized sensitivity function for the toy model presented in Sect. 5.1.4.4. While the raw sensitivity function might suggest that the parameter s_1 is much less important than $k_d 1$, this is due to much larger magnitude of $kd1$. Normalized sensitivity function shows that these parameters are equally important, as their relative changes bring similar relative changes in the system response

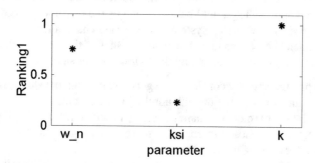

Fig. 6.6 Parameter rankings obtained with local sensitivity analysis for the second order system (6.33)

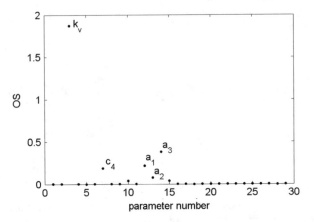

Fig. 6.7 Ranking showing importance of the parameters of NF-κB pathway model for the behavior of the whole system

total parameter ranking, defined by (6.31), is presented in Fig. 6.7. It shows one parameter, k_v, as the highest ranking, significantly more important than any other. It is quite understandable, as this parameter represents the cytoplasmic to nuclear volume ratio [63]. Therefore, its change directly influences all concentrations and therefore all kinetic rates. This parameter should be incorporated in any model treating nucleus and cytosol as separate compartments. However, many models that have been published so far have neglected it. Moreover, it is one of the parameters that can be estimated in a relatively easy way it requires an analysis of the microscopic picture, in which the nuclei of the cells are visible.

However, recalling remarks from Sect. 6.4.3, the biological models refer are based on normalized measurements. In that case, the steady state value (or, more precisely, constant component of the system response) should not affect parameter rankings [65]. Methods based on sensitivity functions become irrelevant then and variance-based approaches should be used. One of the possible methods has been proposed in [65]. As the chosen system responses are characterized by oscillations, the main frequency is chosen to be the characteristics of the system behavior.

The general algorithm is defined by the following four steps:

1. For each parameter under consideration generate its random values (usually from normal or uniform distribution with nominal value as a mean).
2. Run simulation of model dynamics, using nominal values of all parameters but the one that is chosen to come from a randomly generated set. Repeat the simulation for each value from the set.
3. For each simulation run determine oscillation frequency (for simplicity, in this paper it is assumed that there is only one main frequency that carries most energy of the system response). This can be done using, for example, the Discrete Fourier Transform method.
4. Calculate the mean value and variation of calculated frequencies.

The new sensitivity index determining the influence of the p_i on x_j can be now based on the so-called Fano factor [72], defined as

$$\hat{s}_{ij} = \frac{\text{variance}_i}{\text{mean}_i},\tag{6.34}$$

where variance_i and mean_i are variance and mean of the frequency of $x_j(t)$ calculated for simulations in which p_j was varied.

Having applied this algorithm to the system (6.33) we find that the parameter k is not important for the qualitative system behavior (Fig. 6.8), which stands in opposition to the conclusion drawn from analysis based on sensitivity functions.

Similarly, parameter rankings obtained for a toy model of p53 regulatory module presented in Sect. 5.1.4.4, obtained with the sensitivity functions and variance-based method using oscillation frequency appear to be different (Fig. 6.9). This proves

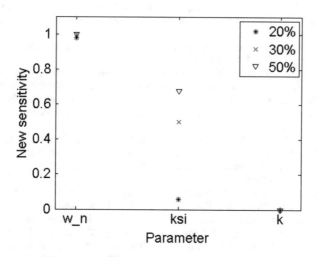

Fig. 6.8 Parameter ranking for the system (6.33) obtained with the frequency-based sensitivity index. Parameter values were sampled from a uniform distribution with $\sigma = 0.2p_n$ (*stars*), $0.3p_N$ (*crosses*), and $0.5p_N$ (*triangles*)

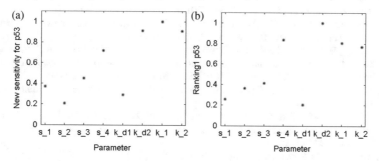

Fig. 6.9 Comparison of parameter rankings for p53 variable based on (**a**) Fano factor and oscillation frequency and (**b**) sensitivity function

that the choice of the sensitivity analysis method should depend on what exactly is known and what the primary focus of such analysis is.

It should be noted that the rankings change with the change of nominal parameter values (Fig. 6.8)—therefore it is crucial to know them with some precision. If they are not known, as is the often case in signaling pathway models, analysis based on sensitivity functions should be supported with other methods.

References

1. I.A. Adzhubei, S. Schmidt, L. Peshkin, V.E. Ramensky, A. Gerasimova, P. Bork, A.S. Kondrashov, S. Sunyaev, A method and server for predicting damaging missense mutations. Nat. Methods **7**, 248–249 (2010)
2. D.G. Albertson, Gene amplification in cancer. Trends Genet. **22**(8), 447–455 (2006)
3. M.L. Avent, V.L. Vaska, B.A. Rogers, A.C. Cheng, S.J. van Hal, N.E. Holmes, B.P. Howden, D.L. Paterson, Vancomycin therapeutics and monitoring: a contemporary approach. Int. Med. J. **43**(2), 110–119 (2013)
4. C. Barnes, Importance of pharmacokinetics in the management of hemophilia. Pediatr. Blood Cancer **60**(Suppl. 1), S27–S29 (2013)
5. C.D. Behrsin, C.J. Brandl, D.W. Litchfield, B.H. Shilton, L.M. Wahl, Development of an unbiased statistical method for the analysis of unigenic evolution. BMC Bioinf. **7**, 150 (2006)
6. M. Bentele, I. Lavrik, M. Ulrich, S. Stosser, D.W. Heermann, H. Kalthoff, P.H. Krammer, R. Eils, Mathematical modeling reveals threshold mechanism in cd95-induced apoptosis. J. Cell Biol. **166**(6), 839–851 (2004)
7. R. Bertolusso, B. Tian, Y. Zhao, L. Vergara, A. Sabree, M. Iwanaszko, T. Lipniacki, A.R. Brasier, M. Kimmel, Dynamic cross talk model of the epithelial innate immune response to double-stranded RNA stimulation: coordinated dynamics emerging from cell-level noise PLoS ONE **9**(4), e93396 (2014)
8. S. Bjorkman, E. Berntorp, Pharmacokinetics of coagulation factors: clinical relevance for patients with haemophilia. Clin. Pharmacokinet. **40**, 815–832 (2001)
9. F.Y. Bois, M. Jamei, H.J. Clewell, PBPK modelling of inter-individual variability in the pharmacokinetics of environmental chemicals. Toxicology **278**(3), 256–267 (2010)
10. P.C. Brown, S.M. Beverly, R.T. Schimke, Relationship of amplified Dihydrofolate Reductase genes to double minute chromosomes in unstably resistant mouse fibroblasts cell lines. Mol. Cell. Biol. **1**, 1077–1083 (1981)
11. Y. Bushkin, F. Radford, R. Pine, A. Lardizabat, B.T. Mangura, M.L. Gennaro, S. Tyagi, Profiling T cell activation using single-molecule fluorescence in situ hybridization and flow cytometry. J. Immunol. **194**(2), 836–841 (2015)
12. F. Campolongo, J. Cariboni, A. Saltelli, An effective screening design for sensitivity analysis of large models. Environ. Model Softw. **22**, 1509–1518 (2007)
13. F.P. Casey, D. Baird, Q. Feng, R.N. Gutenkunst, J.J. Waterfall, C.R. Myers, K.S. Brown, R.A. Cerione, J.P. Sethna, Optimal experimental design in an epidermal growth factor receptor signalling and down-regulation model. IET Syst. Biol. **1**(3), 190–202 (2007)
14. Q. Chang, D. Hedley, Emerging applications of flow cytometry in solid tumor biology. Methods **57**, 359–367 (2012)
15. D. Campbell, O.A. Chkrebtii. Maximum Profile likelihood estimation of differential equation parameters through model based smoothing state estimates bayesian uncertainty. Math. Biosci. **246**(2), 283–292 (2013)
16. J.J. Cruz, *Feedback Systems* (McGraw-Hill, New York, 1972)
17. E. da Fidalgo Silva, S. Botsford, L.A. Porter, Derivation of a novel G2 reporter system. Cytotechnology **68**(1), 19–24 (2016)

18. Z. Darzynkiewicz, H. Crissman, J.W. Jacobberger, Cytometry of the cell cycle: cycling through history. Cytometry A **58A**, 21–32 (2004)
19. S.J. Deminoff, J. Tornow, G.M. Santangelo, Unigenic evolution: a novel genetic method localizes a putative leucine zipper that mediates dimerization of the *Saccharomyces cerevisiae* regulator Gcr1p. Genetics **141**, 1263–1274 (1995)
20. S. Diekmann, C. Hoischen, Biomolecular dynamics and binding studies in the living cell. Phys. Life Rev. **11**(1), 1–30 (2014)
21. M. Dolbniak, M. Kimmel, J. Smieja, Modeling epigenetic regulation of prc1 protein accumulation in the cell cycle. Biol. Direct **10**, 62 (2015)
22. M.J. Downey, D.M. Jeziorska, S. Ott, T.K. Tamai, G. Koentges, et al., Extracting fluorescent reporter time courses of cell lineages from high-throughput microscopy at low temporal resolution. PLoS ONE **6**(12), e2788 (2011)
23. A.D. Fernandes, B.P. Kleinstiver, D.R. Edgell, L.M. Wahl, G.B. Gloor, Estimating the evidence of selection and the reliability of inference in unigenic evolution. Algorithms Mol. Biol. **5**, 35 (2010)
24. L. Ferrante, S. Bompadre, L. Possati, L. Leone, Parameter estimation in a Gompertzian stochastic model for tumor growth. Biometrics **56**(4), 1076–1081 (2000)
25. K. Fujarewicz, M. Kimmel, A. Swierniak, On fitting of mathematical models of cell signaling pathways using adjoint systems. Math. Biosci. Eng. **2**(3), 527–534 (2005)
26. K. Fujarewicz, M. Kimmel, T. Lipniacki, A. Swierniak, Adjoint systems for models of cell signaling pathways and their application to parameter fitting. IEEE/ACM Trans. Comput. Biol. Bioinform. **4**(3), 322–335 (2007)
27. S.N. Gentry, T.L. Jackson, A mathematical model of cancer stem cell driven tumor initiation: implications of niche size and loss of homeostatic regulatory mechanisms PLoS ONE **8**(8), e71128 (2013)
28. M. Girolami, Bayesian inference for differential equations. Theor. Comput. Sci. **408**, 4–16 (2008)
29. R.N. Gutenkunst, F.P. Casey, J.J. Waterfall, C.R. Myers, J.P. Sethna, Extracting falsifiable predictions from sloppy models. Ann. N. Y. Acad. Sci. **1115**(1), 203–211 (2007)
30. R.N. Gutenkunst, J.J. Waterfall, F.P. Casey, K.S. Brown, C.R. Myers, J.P. Sethna, Universally sloppy parameter sensitivities in systems biology models PLoS Comput. Biol. **3**(10), e189 (2007)
31. C.V. Harper, B. Finkenstdt, D.J. Woodcock, S. Friedrichsen, S. Semprini, L. Ashall, D.G. Spiller, J.J. Mullins, D.A. Rand, J.R. Davis, M.R. White, Dynamic analysis of stochastic transcription cycles. PLoS Biol. **9**(4), e1000607 (2011)
32. S. Hicks, D.A. Wheeler, S.E. Plon, M. Kimmel, Prediction of missense mutation functionality depends on both the algorithm and sequence alignment employed. Hum. Mutat. **32**(6), 661–668 (2011)
33. H.C. Ishikawa-Ankerhold, R. Ankerhold, G.P. Drummen. Advanced fluorescence microscopy techniques–FRAP, FLIP, FLAP, FRET and FLIM. Molecules **17**(4), 4047–4132 (2012)
34. R.J. Kaufman, P.C. Brown, R.T. Schimke, Loss and stabilization of amplified dihydrofolate reductase genes in mouse sarcoma S-180 cell lines. Mol. Cell. Biol. **1**, 1084–1093 (1981)
35. K.A. Kim, S.L. Spencer, J.G. Albeck, J.M. Burke, P.K. Sorger, S. Gaudet, H. Kim, Systematic calibration of a cell signaling network model. BMC Bioinf. **11**, 202 (2010)
36. M. Kimmel, D.E. Axelrod, Fluctuation test for two-stage mutations: application to gene amplification. Mutat. Res. **306**, 45–60 (1994)
37. J. Leis, M. Kramer, Sensitivity analysis of systems of differential and algebraic equations. Comput. Chem. Eng. **9**, 93–96 (1985)
38. T. Lipniacki, P. Paszek, A.R. Brasier, B. Luxon, M. Kimmel, Mathematical model of NF-kB regulatory module. J. Theor. Biol. **228**, 195–215 (2004)
39. D. Lu, S. Girish, Y. Gao, B. Wang, J.H. Yi, E. Guardino, M. Samant, M. Cobleigh, M. Rimawi, P. Conte, J.Y. Jin, Population pharmacokinetics of trastuzumab emtansine (T-DM1), a HER2-targeted antibody-drug conjugate, in patients with HER2-positive metastatic breast cancer: clinical implications of the effect of covariates. Cancer Chemother. Pharmacol. **74**(2), 399–410 (2014)

40. A. Marin-Sanguino, S.K. Gupta, E.O. Voit, J. Vera, Biochemical pathway modeling tools for drug target detection in cancer and other complex diseases. Methods Enzymol. **487**, 319–369 (2011)

41. E. Mathe, M. Olivier, S. Kato, C. Ishioka, P. Hainaut, S.V. Tavtigian, Computational approaches for predicting the biological effect of p53 missense mutations: a comparison of three sequence analysis based methods. Nucleic Acids Res. **34**, 1317–1325 (2006)

42. C.A. Miller, B.S. White, N.D. Dees, M. Griffith, J.S. Welch, O.L. Griffith, R. Vij, M.H. Tomasson, T.A. Graubert, M.J. Walter, M.J. Ellis, W. Schierding, J.F. DiPersio, T.J. Ley, E.R. Mardis, R.K. Wilson, L. Ding, SciClone: inferring clonal architecture and tracking the spatial and temporal patterns of tumor evolution PLoS Comput. Biol. **8**, e1003665 (2014)

43. M.D. Morris, Factorial sampling plans for preliminary computational experiments. Technometrics **33**(2), 161–174 (1991)

44. j. Morrow, Genetic analysis of azaguanine resistance in an established mouse cell line. Genetics **65**, 279–287 (1970)

45. J.P. Murnane, M.J. Yezzi, Association of high rate of recombination with amplification of dominant selectable gene in human cells. Somat. Cell Mol. Genet. **14**, 273–286 (1988)

46. G. Neuert, B. Munsky, R.Z. Tan, L. Teytelman, M. Khammash, A. van Oudenaarden, Systematic identification of signal-activated stochastic gene regulation. Science **339**(6119), 584–587 (2013)

47. P.C. Ng, S. Henikoff, Predicting deleterious amino acid substitutions. Genome Res. **11**, 863–874 (2001)

48. H. Nishi, M. Tyagi, S. Teng, B.A. Shoemaker, K. Hashimoto, E. Alexov, S. Wuchty, A.R. Panchenko, Cancer missense mutations alter binding properties of proteins and their interaction networks PLoS ONE **8**(6), e66273 (2013)

49. C.O.T. Oana-Teodora, J.R. Banga, E. Balsa-Canto, Structural identifiability of systems biology models: a critical comparison of methods. PLoS ONE **6**, e27755 (2011)

50. K. Patel, C.M. Kirkpatrick, Pharmacokinetic concepts revisited - basic and applied. Curr. Pharm. Biotechnol. **12**(12), 1983–1990 (2011)

51. T. Peyret, P. Poulin, K. Krishnan, A unified algorithm for predicting partition coefficients for PBPK modeling of drugs and environmental chemicals. Toxicol. Appl. Pharmacol. **249**(3), 197–207 (2010)

52. M. Piazza, X.J. Feng, J.D. Rabinowitz, H. Rabitz, Diverse metabolic model parameters generate similar methionine cycle dynamics. J. Theor. Biol. **251**, 628–639 (2008)

53. K. Puszynski, P. Lachor, M. Kardynska, J. Smieja, Sensitivity analysis of deterministic signaling pathways models. Bull. Pol. Acad. Sci. Tech. Sci. **60**, 471–479 (2012)

54. A. Raj, P. van den Bogaard, S. Rifkin, A. van Oudenaarden, S. Tyagi, Imaging individual MRNA molecules using multiple singly labeled probes. Nat. Methods **5**, 877–887 (2008)

55. V. Ramensky, P. Bork, S. Sunyaev, Human non-synonymous SNPs: server and survey. Nucleic Acids Res. **30**, 3894–3900 (2002)

56. M. Rathinam, P.W. Sheppard, M. Khammash, Efficient computation of parameter sensitivities of discrete stochastic chemical reaction networks. J. Chem. Phys. **132**(3), 034103 (2010)

57. A. Raue, C. Kreutz, T. Maiwald, J. Bachmann, M. Schilling, U. Klingmuller, J. Timmer, Structural and practical identifiability analysis of partially observed dynamical models by exploring the profile likelihood. Bioinformatics **25**, 1923–1929 (2009)

58. B. Reva, Y. Antipin, C. Sander, Predicting the functional impact of protein mutations: application to cancer genomics. Nucleic Acids Res. **39**(17), e118 (2011)

59. M. Roccio, D. Schmitter, M. Knobloch, Y. Okawa, D. Sage, M.P. Lutolf, Predicting stem cell fate changes by differential cell cycle progression patterns. Development **140**(2), 459–470 (2013)

60. A. Rousseau, P. Marquet, Application of pharmacokinetic modelling to the routine therapeutic drug monitoring of anticancer drugs. Fundam. Clin. Pharmacol. **16**(4), 253–262 (2002)

61. A. Saltelli, *Sensitivity Analysis in Practice: A Guide to Assessing Scientific Models* (Wiley, New York, 2004)

62. A. Saltelli, *Global Sensitivity Analysis: The Primer* (Wiley, New York, 2008)

63. J. Smieja, *Dynamics, Feedback Loops and Control in Biology - From Physiological to Individual Cell Models* (Silesian University of Technology, Gliwice, 2011)
64. J. Smieja, M. Jamalludin, A.R. Brasier, M. Kimmel, Model-based analysis of interferon-b induced signaling pathway. Bioinformatics **24**(20), 2363–2369 (2008)
65. J. Smieja, M. Kardynska, A. Jamroz, The meaning of sensitivity functions in signaling pathways analysis. Discrete Contin. Dyn. Syst. Ser. B **10**(8), 2697–2707 (2014)
66. I. Sobol, Global sensitivity indices for nonlinear mathematical models and their monte carlo estimates. Math. Comput. Simul. **55**, 271–280 (2001)
67. R.G. Staudte, R.M. Huggins, J. Zhang, D.E. Axelrod, M. Kimmel, Estimating clonal heterogeneity and interexperiment variability with the bifurcating autoregressive model for cell lineage data. Math. Biosci. **143**(2), 103–121 (1997)
68. E.D. Strome, X. Wu, M. Kimmel, S.E. Plon, Heterozygous screen in *Saccharomyces cerevisiae* identifies dosage-sensitive genes that affect chromosome stability. Genetics **178**(3), 1193–1207 (2008). doi:10.1534/genetics.107.084103
69. M. Sugiyama, A. Sakaue-Sawano, T. Iimura, K. Fukami, T. Kitaguchi, et al., Illuminating cell cycle progression in the developing zebrafish embryo. Proc. Natl. Acad. Sci. **106**, 20812–20817 (2009)
70. D.M. Suter, N. Molina, D. Gatfield, K. Schneider, U. Schibler, F. Naef, Mammalian genes are transcribed with widely different bursting kinetics. Science **332**(6028), 472–474 (2011)
71. S.V. Tavtigian, M.S. Greenblatt, F. Lesueur, G.B. Byrnes, IARC Unclassified Genetic Variants Working Group, In silico analysis of missense substitutions using sequence-alignment based methods. Hum. Mutat. **29**, 1327–1336 (2008)
72. M. Thattai, A. van Oudenaarden, Intrinsic noise in gene regulatory networks. Proc. Natl. Acad. Sci. **98**, 8614–8619 (2001)
73. M.A. Thomas, B. Weston, M. Joseph, W. Wu, A. Nekrutenko, P.J. Tonellato, Evolutionary dynamics of oncogenes and tumor suppressor genes: higher intensities of purifying selection than other genes. Mol. Biol. Evol. **20**(6), 964–968 (2003)
74. T. Tlsty, B.H. Margolin, K. Lum, Differences in the rates of gene amplification in nontumorigenic and tumorigenic cell lines as measured by Luria-Delbruck fluctuation analysis. Proc. Natl. Acad. Sci. **86**, 9441–9445 (1989)
75. A. Tourovskaia, M. Fauver, G. Kramer, S. Simonson, T. Neumann, Tissue-engineered microenvironment systems for modeling human vasculature. Exp. Biol. Med. (Maywood) **239**(9), 1264–1271 (2014)
76. R.D. Travasso, E. Corvera Poir, M. Castro, J.C. Rodrguez-Manzaneque, A. Hernndez-Machado, Tumor angiogenesis and vascular patterning: a mathematical model. PLoS ONE **6**(5), e19989 (2011)
77. A.F. van der Meer, M.A. Marcus, D.J. Touw, J.H. Proost, C. Neef, Optimal sampling strategy development methodology using maximum a posteriori Bayesian estimation. Ther. Drug Monit. **33**(2), 133–146 (2011)
78. A.D. van der Meer, V.V. Orlova, P. ten Dijke, A. van den Berg, C.L. Mummery, Three-dimensional co-cultures of human endothelial cells and embryonic stem cell-derived pericytes inside a microfluidic device. Lab Chip **13**(18), 3562–3568 (2013)
79. N.A.W. Van Riel, Dynamic modelling and analysis of biochemical networks: mechanism based models and model-based experiments. Brief. Bioinform. **7**(4), 364–374 (2006)
80. N.B. Varshaver, M.I. Marshak, N.I. Shapiro, The mutational origin of serum independence in Chinese hamster cells in vitro. Int. J. Cancer **31**, 471–475 (1983)
81. D. Wang, F. Liu, L. Wang, S. Huang, J. Yu, Nonsynonymous substitution rate (Ka) is a relatively consistent parameter for defining fast-evolving and slow-evolving protein-coding genes Biol. Direct **6**, 13 (2011)
82. N. Watanabe, S. Yamashiro, D. Vavylonis, T. Kiuchi, Molecular viewing of actin polymerizing actions and beyond: combination analysis of single-molecule speckle microscopy with modeling, FRAP and s-FDAP (sequential fluorescence decay after photoactivation). Dev. Growth Differ. **55**(4), 508–514 (2013)

83. J.J. Waterfall, F.P. Casey, R.N. Gutenkunst, K.S. Brown, C.R. Myers, P.W. Brouwer, V. Elser, J.P. Sethna, Sloppy-model universality class and the Vandermonde matrix. Phys. Rev. Lett. **97**(15), 150601 (2006)

84. D.J. Woodcock, K.W. Vance, M. Komorowski, G. Koentges, B. Finkenstaedt, D.A. Rand, A hierarchical model of transcriptional dynamics allows robust estimation of transcription rates in populations of single cells with variable gene copy number. Bioinformatics **29**(12), 1519–1525 (2013)

85. X. Wu, E.D. Strome, Q. Meng, P.J. Hastings, S.E. Plon, M. Kimmel, A robust estimator of mutation rates. Mutat. Res. **661**(1–2), 101–109 (2009). doi:10.1016/j.mrfmmm.2008.11.015

86. T.R. Xu, V. Vyshemirsky, A. Gormand, A. von Kriegsheim, M. Girolami, G.S. Baillie, D. Ketley, A.J. Dunlop, G. Milligan, M.D. Houslay, W. Kolch, Inferring signaling pathway topologies from multiple perturbation measurements of specific biochemical species. Sci. Signal. **3**(113), ra20 (2010)

87. T. Yamada, J. Ou, C. Furusawa, T. Hirasawa, T. Yomo, H. Shimizu, Relationship between noise characteristics in protein expressions and regulatory structures of amino acid biosynthesis pathways. IET Syst. Biol. **4**(1), 82–89 (2010)

88. H. Yue, M. Brown, J. Knowles, H. Wang, D.S. Broomhead, D.B. Kell, Insights into the behaviour of systems biology models from dynamic sensitivity and identifiability analysis: a case study of an nf-kappab signalling pathway. Mol. BioSyst. **2**(12), 640–649 (2006)

89. D.E. Zak, R.K. Pearson, R. Vadigepalli, G.E. Gonye, J.S. Schwaber, F.J. Doyle 3rd., Continuous time identification of gene expression models. OMICS: J. Integr. Biol. **7**(4), 373–386 (2003)

90. Y. Zhao, A.R. Brasier, Applications of selected reaction monitoring (SRM)-mass spectrometry (MS) for quantitative measurement of signaling pathways. Methods **61**(3), 313–322 (2013)

91. Z. Zi, K.H. Chob, M.H. Sung, X. Xia, J. Zheng, Z. Sun, In silico identification of the key components and steps in IFN-g induced JAK-STAT signaling pathway. FEBS Lett. **579**, 1101–1108 (2005)

Appendix A
Stability and Controllability of Dynamical Systems

Controllability and stability are essential qualitative properties of dynamical systems. Roughly speaking, controllability generally means, that it is possible to drive a dynamical system from an arbitrary initial state to an arbitrary final state using the set of admissible controls. Stability analysis answers the following question: does the system tend to return to the equilibrium after small perturbations away from it or does it at least remain close to it in some sense?

We are not going to introduce general theory for these two important concepts of mathematical systems theory but we restrict our attention to the class of dynamical systems considered mainly in this book in the context of control theory applications.

To describe this class of systems we use the state equation form of system dynamics characterization.

Definition A.1 A dynamic control system is described by the set of first order differential equations (named state equations):

$$\dot{x}(t) = f\left(x(t), u(t), t\right), \qquad x(t_0) = x_0, \qquad (A.1)$$

where $x(t) \in R^n$ is an n-dimensional state vector, $u(t) \in R^m$ is an m-dimensional input (or control) vector, t_0 is the initial time, and f is an n-dimensional vector function.

If the function f is not dependent explicitly on time t, then we say that the system is time-invariant and its description (A.1) may be written as

$$\dot{x}(t) = f\left(x(t), u(t)\right), \qquad x(t_0) = x_0. \qquad (A.2)$$

© Springer International Publishing Switzerland 2016
A. Świerniak et al., *System Engineering Approach to Planning Anticancer Therapies*, DOI 10.1007/978-3-319-28095-0

For simplicity in notation we will omit (when it does not lead to any ambiguity) the dependence of state and control vectors on t, which leads to the model in the form:

$$\dot{x} = f(x, u), \qquad x(t_0) = x_0. \tag{A.3}$$

Moreover, in the case of $u \equiv 0$, the system is called unforced (or uncontrolled, or autonomous) and we will write simply

$$\dot{x} = f(x), \qquad x(t_0) = x_0. \tag{A.4}$$

Although notions of stability and controllability could be defined for the system (A.1), we will reduce our attention to the system (A.3) for controllability and (A.4) for stability analysis to make the things as simple as possible for understanding of the topics discussed in the book. Moreover, we assume that the function f is endowed with such properties that guarantee existence and uniqueness of solution of the state equation, given an initial condition and a control vector.

To start the analysis of system stability we consider equilibrium points of the autonomous unforced system (A.4), defined as solutions of the algebraic equation:

$$f(x_e) = 0. \tag{A.5}$$

Consider perturbations about the equilibrium points $x = x_e + \bar{x}$:

Definition A.2 A linear system

$$\dot{\bar{x}} = A\bar{x} \tag{A.6}$$

is called an associated linear system for system (A.3) near the equilibrium x_e if $A = \left(\frac{\partial f}{\partial x}\right)\Big|_{x_e}$ where A is an $n \times n$ matrix.

Probably the easiest way to examine the linearized model stability around the equilibrium points is through calculation of the determinant and trace of the matrix A. The system is stable at a given equilibrium only if the determinant is positive and trace negative. Figure A.1 shows how the stability and type of the equilibrium depends on the matrix A determinant and trace.

For simplicity, let us assume that the equilibrium point is at the space origin. There may exist many equilibrium points, nevertheless we may assume that the equilibrium point is at the origin, since if $x_e \neq 0$, a simple translation can always be applied to obtain an equivalent system with the equilibrium at 0.

Definition A.3 System (A.4) is stable (in the sense of Lyapunov) in the neighborhood of its equilibrium at the origin if for any $\epsilon > 0$, there exist $\delta(\epsilon, t_0)$ such that if $\|x(t_0)\| < \delta$, then $\|x(t)\| < \epsilon$ for any $t > t_0$. If δ does not depend on t_0, then the system is uniformly stable around the equilibrium [9].

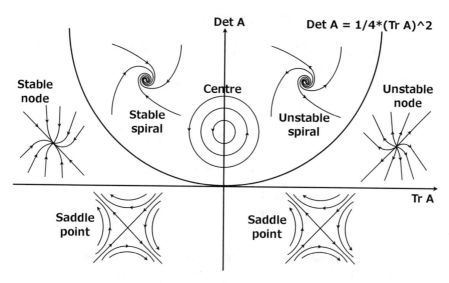

Fig. A.1 Equilibrium points types and their stability

If the system is stable and additionally

$$\lim_{t \to \infty} x(t) = 0 \qquad \text{for any} \qquad x_0 \qquad \text{such that} \qquad \|x(t_0)\| < \delta \qquad \text{(A.7)}$$

then the system is asymptotically stable around the equilibrium.

Definition A.4 The system is globally asymptotically stable (asymptotically stable in the large) if it is asymptotically stable and (A.7) is satisfied for any initial condition x_0.

The local asymptotic stability (in the small neighborhood of the equilibrium point) could be checked by the use of linearization method (Lyapunov's first method).

Theorem A.1 (Lyapunov's First Method [10]) *If all eigenvalues λ_i of matrix A (roots of the characteristic equation: $\det(A - \lambda I) = 0$) of the linear system (A.6) associated with the system (A.3) near the origin have negative real parts, then the origin is a locally asymptotically stable equilibrium of (A.3). If there exists an eigenvalue of A with positive real part, then the origin is locally unstable. If all eigenvalues have negative real parts except for some whose real parts are equal to zero, then the method does not allow to determine local stability.*

Recall that eigenvectors v and eigenvalues λ of a matrix A are related by the equation: $Av = \lambda v$.

It should be noted that, once the matrix A is known, one does not need to calculate the eigenvalues to check system stability. Instead, appropriate stability criterion can be utilized, for example Hurwitz criterion.

Let us introduce the characteristic equation of the system (A.3):

$$(\lambda I - A) = a_0\lambda^n + a_1\lambda^{n-1} + \cdots + a_{n-1}\lambda + a_n = 0, \tag{A.8}$$

where I is the identity matrix. Then the $n \times n$ matrix

$$H = \begin{bmatrix} a_1 & a_3 & a_5 & a_7 & \ldots & 0 \\ a_0 & a_2 & a_4 & a_6 & \ldots & 0 \\ 0 & a_1 & a_3 & a_5 & \ldots & 0 \\ 0 & a_0 & a_2 & a_4 & \ldots & 0 \\ 0 & 0 & a_1 & a_3 & \ldots & 0 \\ \vdots & \vdots & \vdots & \vdots & \ddots & \vdots \\ 0 & 0 & 0 & 0 & \ldots & a_n \end{bmatrix} \tag{A.9}$$

is called the Hurwitz matrix corresponding to the characteristic equation (A.8). In [5] Hurwitz showed that eigenvalues of the A matrix had strictly negative real parts (and, consequently, the system described by the A matrix will be asymptotically stable) if and only if all the leading principal minors of the matrix $H(\lambda)$ were positive:

$$\Delta_i = \begin{vmatrix} a_1 & a_3 & \ldots \\ a_0 & a_2 & \ldots \\ \vdots & & \\ 0 & \ldots & a_n \end{vmatrix} > 0 \tag{A.10}$$

for $i = 1, 2, \ldots, n$.

One of the most often used algorithms to check these dependencies is the Routh's algorithm (see, e.g., [3]).

It should be noted that the method based on model linearization fails to provide a definite answer on system stability if the eigenvalues of matrix A are purely imaginary. Moreover, it does not provide any information about the basin of attraction of the equilibrium point. These problems can be dealt with using the so-called Lyapunov's direct method.

Definition A.5 Let V be a continuously differentiable function of x (mapping R^n into R). We call $V(x)$ positive definite in some region Ω containing the origin if

1. $V(0) = 0$.
2. $V(x) > 0$ for any x in this region except for the origin.

If inequality (2) is weakly satisfied, i.e. $V(x) \geq 0$, then we say that V is positive semidefinite. Similarly, if we change the sign in the inequality, we have negative definite or negative semidefinite functions, respectively.

The derivative of V with respect to time along a trajectory of the system (A.3) denoted by $\dot{V}(x(t))$ is given by

$$\dot{V}(x(t)) = \frac{dV}{dx}\dot{x} = \frac{dV}{dx}f(x). \tag{A.11}$$

Definition A.6 Let V be a positive definite function in some region Ω. If its derivative along a trajectory $\dot{V}$ is negative semidefinite in some region Ω_1 around the origin, then V is a Lyapunov function.

Theorem A.2 *If there exists a Lyapunov function for the system (A.3), then the origin is a stable equilibrium point. If, in addition, $\dot{V}$ is negative definite in some region around the origin, then the system (A.3) is asymptotically stable near the origin.*

Theorem A.3 (Lyapunov's Direct Method) *If there exists a Lyapunov function V over the entire state space with derivative $\dot{V}$ negative definite in the entire space and V is radially unbounded, i.e. $V(x(t)) \rightarrow \infty$ if $\|x(t)\| \rightarrow \infty$, then the system is globally asymptotically stable (the origin is asymptotically stable in the large).*

The name "direct" of this method is due to the fact that stability can be determined without explicitly solving the system equations. It is also called the second Lyapunov's method. In general, the hardest condition to meet in Lyapunov's second method is to find a V function that yields $\dot{V} < 0$. What happens quite often is that a given V yields $\dot{V} \leq 0$ (the system is stable, instead of asymptotically stable). A very useful theorem, due to La Salle, called the invariance principle, can often be used in this case.

Theorem A.4 (La Salle Invariance Principle [10]) *Let $V : \quad R^n \rightarrow R$ be a locally positive definite function such that on the compact set $\Omega_c = \{x \in R^n : \quad V(x) \leq c\}$ we have $\dot{V}(x) \leq 0$. Define*

$$S = x \in \Omega_c : \dot{V} = 0$$

As $t \rightarrow \infty$, the trajectory tends to the largest invariant set inside S. In particular, if S contains no invariant sets other than $x = 0$, then 0 is asymptotically stable.

If there exists a Lyapunov function V over the entire state space with derivative $\dot{V}$ negative semidefinite in the entire space and V is radially unbounded, and $\dot{V} \neq 0$ along any solution of the differential equation except the origin, then the system is globally asymptotically stable.

In our considerations related to controllability of systems we also restrict our attention to time invariant systems (A.2), but since the essential question deals with control variables, only state equations containing control equations are interesting for us.

Definition A.7 The dynamical system (A.2) is said to be controllable if for every initial condition x_0 and every vector $x_1 \in R^n$ there exist a finite time t_1 and control $u(t) \in R^m$, such that $x(t_1; x_0, u(t)) = x_1$ (e.g., [7]).

Generally, finding conditions of controllability for nonlinear system is difficult and depends on the type of function in the state equation, a set of initial and final states considered and the control interval $[t_0, t_1]$.

However, in the case of linear systems described by the linear state equations:

$$\dot{x} = Ax + Bu, \qquad\qquad x(t_0) = x_0, \qquad\qquad (A.12)$$

where A is an $n \times n$ state matrix and B is an $n \times m$ input matrix, the following rank condition was formulated by Kalman in the 1960s of the previous century [6].

Theorem A.5 (Rank Condition) *The system (A.12) is completely controllable if and only if the $n \times nm$ block matrix [6]: $Q = [B|AB|A^2B| \ldots |A^{n-1}B]$ has rank equal to n.*

It is worth to note that the condition is both necessary and sufficient and it is independent of initial and final conditions, and of the time interval in which the system is controlled.

In practice admissible controls are always required to satisfy certain additional constraints, in this case we may discuss constrained controllability. Generally, for arbitrary control constraints it is rather very difficult to give easily computable criteria for constrained controllability. However, for some special cases of the constraints it is possible to formulate and prove simple algebraic constrained controllability conditions. Therefore, we assume that the set of values of controls $U_c \subset U(R^m)$ is a given closed and convex cone with nonempty interior and vertex at zero. In the sequel we shall focus our attention on the so-called constrained controllability in the time interval $[t_0, t_1]$. In order to do that, first of all let us introduce the notion of the attainable set at time $t_1 > t_0$ from zero initial conditions, denoted shortly by $K_T(U_c)$ and defined as follows:

$$K_T(U_c) = \{x(t) \in X(R^n) : \quad x = x(t_1; u), \quad u(t) \in U_c \text{ for a.e.} t \in [t_0, t_1]\} \tag{A.13}$$

where $x(t; u)$, $t > 0$ is the unique solution of the differential state equation (A.2) with zero initial conditions and a given admissible control.

Now, using the concept of the attainable set, we may formulate definitions of the constrained controllability in $[t_0, t_1]$.

Definition A.8 The dynamical system (A.2) is said to be U_c-locally controllable in $[t_0, t_1]$ if the attainable set $K_T(U_c)$ contains a neighborhood of zero in the space X.

Definition A.9 The dynamical system (A.2) is said to be U_c-globally controllable in $[t_0, t_1]$ if $K_T(U_c) = X$.

For linear system (A.12) we have the following condition of constrained controllability:

Theorem A.6 *Suppose the set U is a cone with vertex at zero and a nonempty interior in the space R^m. Then, the dynamical system (A.12) is U-controllable from zero if and only if [2]*

1. *The rank of the block matrix $Q = [B|AB|A^2B| \ldots |A^{n-1}B]$ is equal to n.,*
2. *There is no real eigenvector $v \in R^n$ of the transposed matrix A^T satisfying $v^T Bw \le 0$ for all $w \in U$*

If, additionally, values of control variables are bounded from above, then to drive the system to the neighborhood of zero the matrix A should have no eigenvalues with positive real parts.

In the special case for the single input system i.e., $m = 1$, condition (2) of Theorem A.6 reduces to the requirement that matrix A has no real eigenvalues.

For more in-depth description of the concepts of stability and controllability, the reader might turn to classical textbooks, e.g., [1, 8, 11, 12] and for methods tailored specifically to nonnegative systems, which are the subject of this book, to, e.g., [4].

References

1. A. Bacciotti, L. Rosier, *Liapunov Functions and Stability in Control Theory* (Springer, Berlin, 2005)
2. R.F. Brammer, Controllability in linear autonomous systems with positive controllers. SIAM J. Control **10**(2), 339–353 (1972)
3. G. Goodwin, S. Graebe, M. Salgado, *Control System Design* (Prentice Hall, Upper Saddle River, 2000)
4. W.M. Haddad, V.S. Chellaboina, Q. Hui, *Nonnegative and Compartmental Dynamical Systems* (Princeton University Press, Princeton, 2010)
5. A. Hurwitz, Ueber die Bedingungen, unter welchen eine Gleichung nur Wurzeln mit negativen reellen Teilen besitzt [About the conditions under which an equation has only roots with negative real parts]. Math. Ann. **46**, 273–284 (1895)
6. R.E. Kalman, On the general theory of control systems, in *Proceeding of the First IFAC Congress Automatic Control*, Moscow, vol. 1 (Butterworths, London, 1960), pp. 481–492
7. J. Klamka, *Controllability of Dynamical Systems* (Kluwer Academic, Dordrecht, 1991)
8. W. Krabs, W. Prickl, *Dynamical Systems. Stability, Controllability and Chaotic Behavior* (Springer, Berlin, 2010)
9. A.M. Lyapunov, The general problem of the stability of motion (in Russian). Doctoral dissertation, University Kharkov, 1892; English translation: Stability of Motion (Academic, New York/London, 1966)
10. J.L. Salle, S. Lefschetz, *Stability by Liapunov's Direct Method* (Academic, New York, 1961)
11. T.L. Vincent, W.J. Grantham, *Nonlinear and Optimal Control Systems* (Wiley, New York, 1997)
12. V.I. Vorotnikov, *Partial Stability and Control* (Birkhauser, Boston, 1998)

Appendix B
Pontryagin Maximum Principle and Optimal Control

Pontryagin Maximum Principle is one of the most powerful tools for optimization of dynamical control systems. Although it formulates only necessary conditions of optimality of control variables and related trajectories, these conditions may serve as a basis for complete synthesis of optimal solutions to practical dynamic optimization problems. The conditions are particularly useful in the case of time-continuous finite dimensional dynamic systems, when the Maximum Principle gives global and strong necessary conditions of optimality for a broad class of admissible controls including measurable functions and generally formulated constraints imposed on control and state variables. Moreover, there exist many extensions of Pontryagin Maximum Principle for infinite dimensional systems. However, they are usually weaker than the basic version.

Taking into account the applications of the Maximum Principle to optimization problems presented in this book, similarly as in Appendix A we restrict our consideration to the class of systems described by the finite set of state equations (A.1) with control variables being piecewise continuous functions of time. Moreover, the conditions will be discussed for time-invariant systems, considered in this book, as it was presented in the original work of Pontryagin and co-workers [6].

The optimal control problem may be formulated as follows: Given a dynamical control system (A.1) with f continuous in (x, u, t), differentiable in x for fixed (u, t), and such that the partial derivatives of f with respect to x are continuous as a function of all variables. The admissible controls are piecewise continuous functions $u(t) \subset R$ defined on a compact interval $[t_0, t_1]$ with values in the control set U. The objective, or cost functional (also called a performance index), is given as

$$J(u) = \int_{t_0}^{T} L\left(x(t), u(t), t\right) dt + \Phi\left(x(T), T\right), \tag{B.1}$$

© Springer International Publishing Switzerland 2016
A. Świerniak et al., *System Engineering Approach to Planning
Anticancer Therapies*, DOI 10.1007/978-3-319-28095-0

where x is the unique trajectory (unique solution of (A.1)) corresponding to the control u. L is usually called Lagrangian and is endowed with the same properties as f. Φ is a continuously differentiable penalty function. The terminal time $T \leq t_1$ can be fixed or free. Terminal state $x(T)$ can be free, constrained by some terminal target set, or fixed. In the second case we assume that the target set is defined by the system of k equations:

$$\Psi(x(T), T) = 0 \qquad (B.2)$$

and in the third case there is no point in including the penalty function Φ in the cost functional.

The optimal control problem is to minimize the objective $J(u)$ over all admissible controlled trajectories (x, u) defined over an interval $[t_0, T]$ that satisfy the possible terminal constraints.

As mentioned before we restrict our consideration to time-invariant problems and similarly as in (A.1) in notation we omit dependence of state and control variables on time (when it does not lead to ambiguity).

Thus, in further considerations the system will be described by the state equation (A.3) and the cost functional will be reduced to

$$J(u) = \int_{t_0}^{T} L(x, u)dt + \Phi(x(T)). \qquad (B.3)$$

To formulate conditions of optimality we first introduce the so-called Hamiltonian function and adjoint variables.

Definition B.1 (Hamiltonian) The Hamiltonian function $H(x, u, p_0, p)$ is given by

$$H(x, u, p_0, p) = p_0 L(x, u) + p^T f(x, u), \qquad (B.4)$$

where p_0 is a constant scalar multiplier and $p(t)$ is a vector of time dependent multipliers, called also adjoint variables or co-state variables.

The necessary conditions of optimality could be now formulated by the following theorem.

Theorem B.1 (Pontriagin Maksimum Principle) *Let (x^*, u^*) be a controlled trajectory defined over the interval $[t_0, T]$ with the control u^* piecewise continuous. If (x^*, u^*) is optimal, then there exist a constant $p_0 \geq 0$ and adjoint variables $p(t) \in R^n$ not equal to zero simultaneously (nontriviality condition), such that*

- *The adjoint variables p satisfy the adjoint differential equation*

$$\dot{p} = -\left(\frac{\partial H}{\partial x}\right)^T \qquad (B.5)$$

- *The Hamiltonian satisfies the minimum condition:*

$$H(x^*, u^*, p_0, p) = \min_{u \in U} H(x^*, u, p_0, p) = \text{const.} \qquad (\text{B.6})$$

If both terminal time and terminal state are fixed, there are no other conditions. If the terminal state is free, then the adjoint vector satisfies a transversality condition at the terminal time:

$$p(T) = \left(\frac{\partial \Phi}{\partial x} \right)^T. \qquad (\text{B.7})$$

If the terminal state is constrained by the target set defined by (B.2), then the transversality condition requires existence of a k-dimensional vector μ of constant multipliers such that

$$p(T) = \left(\frac{\partial \Phi}{\partial x} \right)^T + \mu^T \left(\frac{\partial \Psi}{\partial x} \right). \qquad (\text{B.8})$$

If the terminal time is free, then, additionally to previously stated conditions, the Hamiltonian H vanishes identically along the optimal controlled trajectory, i.e.

$$H(x^*, u^*, p_0, p) = \min_{u \in U} H(x^*, u, p_0, p) = 0. \qquad (\text{B.9})$$

Thus, regardless of the conditions imposed on terminal state, the maximum principle leads to the two point boundary value problem (TPBVP) for state and co-state variables, augmented by the minimization condition for control variables. The non-triviality condition precludes solution in which $(p_0, p) = (0, 0)$. The solution in which $p_0 > 0$ is called normal and when $p_0 = 0$, abnormal. Since the Hamiltonian is linear in p_0, p, in the normal case we may simply scale them and take $p_0 = 1$. That is the typical case for many practical optimization problems.

The use of Maximum Principle requires minimization of Hamiltonian with respect to control variables. The problem begins when the Hamiltonian function does not depend explicitly on u on some finite interval of time. It leads to the so-called singular optimal control problems [3] and in this case higher order necessary conditions of optimality should be applied. For problems with functions f and L twice differentiable with respect to u, such case appears when Hessian matrix $\frac{\partial^2}{\partial u^2} H$ is singular. Sometimes this condition is treated as a definition of singular control. It is worth to note that for singular optimal control in scalar case the condition $\frac{\partial H}{\partial u}$ is trivially satisfied for any admissible control in the finite time horizon.

The singularity of optimal control is a problem which should be especially taken into account when Hamiltonian linearly depends on control (is an affine function of control variables). In other words, functions f and L are defined as follows:

$$f(x, u) = f_0(x) + B(x)u,$$
$$L(x, u) = l(x) + G(x)u$$

(B.10)

In this case:

$$\frac{\partial H}{\partial u} = p_0 G(x) + p^T B(x)$$

(B.11)

and does not depend on u. The sign of this derivative determines the minimum of Hamiltonian. Change in sign implies switching of the control variable from its minimum to maximum value. Thus (B.11) is called a switching function S and control strategy based on switching between its minimum to maximum is called a bang-bang control. If, however, in the finite time interval $I \subset [t_0, T]$, S stays equal to zero, then we are led to the problem of singular controls and relative trajectories called singular arcs. Singular controls could be calculated by differentiating the switching functions with respect to time until the control variable explicitly appears in the derivative. It is known that the number of such successive differentiation must be even. Denote it by $2q$.

Definition B.2 The smallest integer (if any) q such that

$$\frac{\partial}{\partial u} \frac{d^{2q}}{dt^{2q}} \frac{\partial H}{\partial u} \neq 0$$

(B.12)

on interval I is called the order of singular arc.

This definition enables formulation of the necessary condition for the optimality of a singular arc:

Theorem B.2 (Generalized Legendre–Clebsch Condition)
   If a singular arc of order q is optimal, then on interval I:

$$\frac{\partial}{\partial u} \frac{d^{2q}}{dt^{2q}} \frac{\partial H}{\partial u} \geq 0$$

(B.13)

on this interval.

This condition can be used for single input systems (with a scalar control) or component-wise for multi input systems (with many control variables). But if many control variables are simultaneously singular, then other higher order necessary conditions should be applied. For example, in the case of two control variables u_1, u_2 the following necessary condition can be formulated.

Theorem B.3 (Goh Condition) *If singular controls u_1, u_2 are singular on the same interval I, then*

$$\frac{\partial}{\partial u_2} \frac{d}{dt} \frac{\partial H}{\partial u_1} = 0 \tag{B.14}$$

identically on this interval.

It is worth noting that the term $\frac{\partial H}{\partial u}$, present in all conditions, represents the switching function S for the considered problems.

For more in-depth description of the Pontryagin Maximum Principle and optimal control, the reader might turn to classical textbooks, e.g., [1, 2, 4, 5]. Examples of applications of optimal control theory in biomedical field can be found in [7].

References

1. L.D. Berkovitz, N.G. Medhin, *Nonlinear Optimal Control Theory* (Chapman and Hall/CRC, Boca Raton, 2012)
2. V.G. Boltyanski, A. Poznyak, *The Robust Maximum Principle. Theory and Applications* (Birkhauser, Boston, 2012)
3. B. Bonnard, M. Chyba, *Singular Trajectories and Their Role in Control Theory* (Springer, Berlin, 2003)
4. G. Knowles (ed.), *An Introduction to Applied Optimal Control* (Academic Press, New York, 1981)
5. D.A. Pierre, *Optimization Theory with Applications* (Dover Publications, New York, 2012)
6. L. Pontryagin, V. Boltyanski, R. Gamkrelidze, E. Mischenko, *Mathematical Theory of Optimal Processes* (Wiley, New York, 1962)
7. H. Schaettler, U. Ledzewicz, *Geometric Optimal Control. Theory, Methods and Examples* (Springer, New York, 2012)

Appendix C
Bifurcation Analysis

The term *Bifurcation* was introduced by Henri Poincaré in 1885 [2]. System bifurcation occurs when a continuous change of one or more of its parameters causes qualitative change of system behavior. These transitions may be summarized as follows:

stable	↔	unstable
symmetric	↔	asymmetric
stationary	↔	motion
regular	↔	irregular
order	↔	chaos

One of the main bifurcation characteristics is codimension which describes the number of the parameters that have to change simultaneously to cause systems bifurcation. The formal description of the most common codimension one bifurcation in one-dimensional systems can be found in [3] and [1].

Definition of the codimension one bifurcation in higher-dimensional systems can be performed by analysis of the Jacobian matrix. Let us assume that the system is described by following equation system:

$$\dot{x} = f(x, t). \tag{C.1}$$

Then, quadratic matrix:

$$J = \frac{\partial f}{\partial x} = \begin{bmatrix} \frac{\partial f_1}{\partial x_1} & \frac{\partial f_1}{\partial x_2} & \cdots & \frac{\partial f_1}{\partial x_n} \\ \frac{\partial f_2}{\partial x_1} & \frac{\partial f_2}{\partial x_2} & \cdots & \frac{\partial f_2}{\partial x_n} \\ \vdots & \vdots & \ddots & \vdots \\ \frac{\partial f_n}{\partial x_1} & \frac{\partial f_n}{\partial x_2} & \cdots & \frac{\partial f_n}{\partial x_n} \end{bmatrix} \tag{C.2}$$

© Springer International Publishing Switzerland 2016
A. Świerniak et al., *System Engineering Approach to Planning Anticancer Therapies*, DOI 10.1007/978-3-319-28095-0

is called the Jacobian matrix. Special form of this matrix is the Jacobian matrix at point P received by linearization of Eq. (C.1) around a trajectory, most commonly around systems equilibrium.

System linearization is made by expanding the functions in a Taylor series with only first order terms retained:

$$\sum_{i=1}^{N} \left(\frac{\partial f(t,x)}{\partial x_i} \bigg|_{P} \Delta x_i \right) \tag{C.3}$$

We receive the system of equations of the form:

$$\Delta \dot{x}(t) = J_P \Delta x(t) \tag{C.4}$$

where J_P is Jacobian matrix at the equilibrium. Determinant and trace of this matrix indicate stability and type of the equilibrium point P. Bifurcation occurs when the system changes its equilibrium points type and/or number, or the equilibria collide with the change of the bifurcation parameter. Below, based on [3], we will briefly describe most common codimension one bifurcations. Extended mathematical description of this bifurcation types and its conditions can be found in [3].

C.1   Flip Bifurcation

Flip bifurcation is also called period doubling bifurcation [3, p. 28]

Let $f(x, \mu) : R^2 \rightarrow R$ be a one-parameter family of C^3 maps satisfying

$$f(0,0) = 0,$$

$$\frac{\partial f}{\partial x} \bigg|_{x=0, \mu=0} = -1,$$

$$\frac{\partial^2 f}{\partial x \partial \mu} \bigg|_{x=0, \mu=0} < 0,$$

$$\frac{\partial^3 f}{\partial x^3} \bigg|_{x=0, \mu=0} < 0 \tag{C.5}$$

Then there are intervals $(\mu_1, 0)$ and $(0, \mu_2)$ and $\epsilon > 0$ such that

- if $\mu \in (\mu_1, 0)$, then $f_\mu(x)$ has a single fixed stable point for $x \in (-\epsilon, \epsilon)$,
- if $\mu \in (0, \mu_2)$, then for $x \in (-\epsilon, \epsilon)$, $f_\mu(x)$ has one stable orbit of period 2 and one unstable fixed point.

A fixed point x_f is defined in [3] (Eq. (2.5)) as point satisfying

$$f(x_f) = x_f \tag{C.6}$$

An example of a system exhibiting flip bifurcation is a system described by equation:

$$f(x, \mu) = \mu - x - x^2 \qquad (C.7)$$

C.2   Fold Bifurcation

Fold bifurcation is also called saddle node bifurcation [3, p. 27].

For the one-dimensional systems let $f(x, \mu) : R^2 \to R$ be a one-parameter family of C^2 maps satisfying

$$f(0, 0) = 0,$$

$$\left. \frac{\partial f}{\partial x} \right|_{x=0, \mu=0} = 1,$$

$$\left. \frac{\partial^2 f}{\partial x^2} \right|_{x=0, \mu=0} > 0,$$

$$\left. \frac{\partial f}{\partial \mu} \right|_{x=0, \mu=0} > 0 \qquad (C.8)$$

Then there exist intervals $(\mu_1, 0)$ and $(0, \mu_2)$ and $\epsilon > 0$ such that

- if $\mu \in (\mu_1, 0)$, then for $x \in (-\epsilon, \epsilon)$, $f_\mu(x)$ has two fixed points: one positive being unstable and the second negative and stable,
- if $\mu \in (0, \mu_2)$, then $f_\mu(x)$ has no fixed points for $x \in (-\epsilon, \epsilon)$

Simple one-dimensional system exhibiting fold bifurcation is described by equation:

$$\dot{x} = \mu + x + x^2 \qquad (C.9)$$

For the multidimensional systems this type of bifurcation occurs when the system has at least two equilibium points with one node and the second saddle point type. A saddle-node bifurcation is a collision and disappearance of these two equilibria. This occurs when the critical equilibrium has one zero eigenvalue. This means that with the increase of the bifurcation parameter both equilibria, node and saddle, change their position and become one double critical equilibrium at bifurcation point with the determinant of J_P equal to 0, caused by one eigenvalue equal to 0. With a further bifurcation parameter increase both equilibrium points lose their real parts and disappear (Fig. C.1).

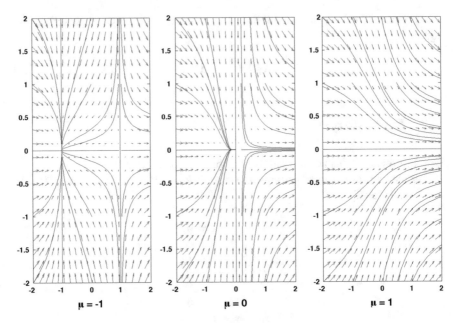

Fig. C.1 Two-dimensional saddle-node bifurcation example

A simple multidimensional system exhibiting fold bifurcation is described by equations:

$$\dot{x} = \mu + x + x^2$$
$$\dot{y} = -y \qquad\qquad\qquad (C.10)$$

C.3   Pitchfork Bifurcation

For the one-dimensional systems let $f(x, \mu) : R^2 \rightarrow R$ be a one-parameter family of C^3 maps satisfying

$$f(-x, \mu) = -f(x, \mu),$$

$$\left.\frac{\partial f}{\partial x}\right|_{x=0,\mu=0} = 1,$$

$$\left.\frac{\partial}{\partial \mu}\left(\left.\frac{\partial f}{\partial x}\right|_{x=0}\right)\right|_{x=0,\mu=0} > 0,$$

$$\left.\frac{\partial^3 f}{\partial x^3}\right|_{x=0,\mu=0} < 0 \qquad\qquad (C.11)$$

Then there exist intervals $(\mu_1, 0)$ and $(0, \mu_2)$ and $\epsilon > 0$ such that

- if $\mu \in (\mu_1, 0)$, then $f_\mu(x)$ has a single stable fixed point at in $(-\epsilon, \epsilon)$,
- if $\mu \in (0, \mu_2)$, then $f_\mu(x)$ has three fixed points in $(-\epsilon, \epsilon)$. The origin $x = 0$ is unstable while two other are stable.

Simple example of one-dimensional system exhibiting supercritical pitchfork bifurcation [3, p. 31] is given by formula:

$$\dot{x} = \mu x + x - x^3 \tag{C.12}$$

For the multidimensional systems pitchfork bifurcation is the type of the bifurcation where the number of equilibrium points and their type changes. This is the result of the quadratic form of the equilibrium term crossing through 0 with a simultaneously change of the sign of Jacobian matrix determinant. We distinguish two types of this bifurcation: supercritical and subcritical.

In supercritical pitchfork bifurcation as the bifurcation parameter increases one stable node changes its type to the saddle with simultaneous birth of two additional node type equilibria (Fig. C.2).

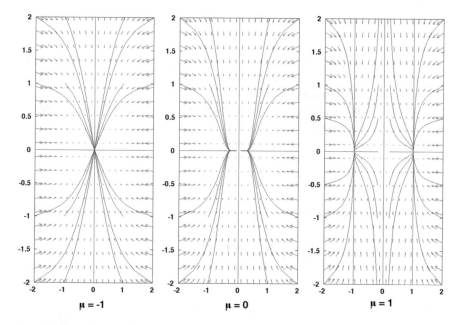

Fig. C.2 Two-dimensional supercritical pitchfork bifurcation example

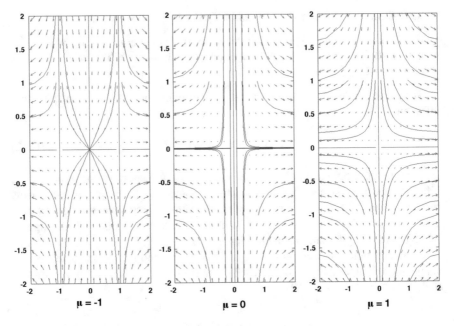

Fig. C.3 Two-dimensional subcritical pitchfork bifurcation example

In the subcritical pitchfork bifurcation with the bifurcation parameter increase, two saddle type and one node type equilibrium points collide and collapse to form one saddle type equilibrium point (Fig. C.3).

A simple example of a supercritical Pitchfork bifurcation in the multidimensional system is given by formula:

$$\dot{x} = \mu x + x - x^3$$
$$\dot{y} = -y \qquad\qquad\qquad (C.13)$$

while a subcritical one is given by

$$\dot{x} = \mu x + x + x^3$$
$$\dot{y} = -y \qquad\qquad\qquad (C.14)$$

C.4   Transcritical Bifurcation

For the one-dimensional systems let $f(x, \mu) : R^2 \rightarrow R$ be a one-parameter family of C^2 maps satisfying

$$f(0, \mu) = 0,$$

$$\left.\frac{\partial f}{\partial x}\right|_{x=0,\mu=0} = 1,$$

$$\left.\frac{\partial}{\partial \mu} \left(\left.\frac{\partial f}{\partial x}\right|_{x=0} \right) \right|_{x=0,\mu=0} > 0,$$

$$\left.\frac{\partial^2 f}{\partial x^2}\right|_{x=0,\mu=0} > 0 \qquad \text{(C.15)}$$

Then there are two branches, of which one is stable and second is unstable.

An example of this type of bifurcation in a one-dimensional system is given by

$$\dot{x} = \mu x + x + x^2 \qquad \text{(C.16)}$$

For the multidimensional systems transcritical bifurcation occurs when at least two equilibrium points, node and saddle type, change their position and type as the determinant of the Jacobian matrix crosses 0 [3, p. 30]. With the bifurcation parameter increase equilibria move to each other and at the critical value they form one double equilibrium. At this point one eigenvalue of the Jacobian matrix is equal to 0 and therefore the determinant of the J_P is equal to 0. With a further increase of the bifurcation parameter equilibrium points move away from each other and the node type equilibrium becomes saddle while saddle type becomes node (Fig. C.4).

An example of this type of bifurcation in a multidimensional system is given by:

$$\dot{x} = \mu x + x + x^2$$

$$\dot{y} = -y \qquad \text{(C.17)}$$

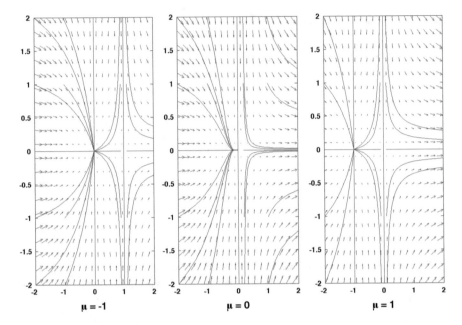

Fig. C.4 Two-dimensional transcritical bifurcation example

C.5   Hopf Bifurcation

Hopf bfurcation is also called the Andronov-Hopf bifurcation.

First we have to define the limit cycle. Following [3] (p. 218) we can define it as "an attracting set to which orbits or trajectories converge and upon which the dynamics is periodic." Then for the multidimensional system Hopf bifurcation is the ascendance of a limit cycle from an equilibrium in dynamical system, when the equilibrium changes stability through a pair of purely imaginary eigenvalues. In other words, the trace of the J_P matrix crosses 0 while its determinant becomes positive as the real parts of the eigenvalues change their sign.

There are four main types of codimension one bifurcations leading to oscillations: supercritical Hopf (superH), subcritical Hopf (subH), saddle-node-on-invariant-circle (SNIC), and saddle-loop (SL)

These four types differ in the amplitude and period of the oscillations:

- In supercritical Hopf bifurcation, with the increase of the bifurcation parameter a stable spiral becomes unstable, while simultaneously a stable limit-cycle emerges (Fig. C.5); oscillations arise with zero amplitude and finite (nonzero) period.

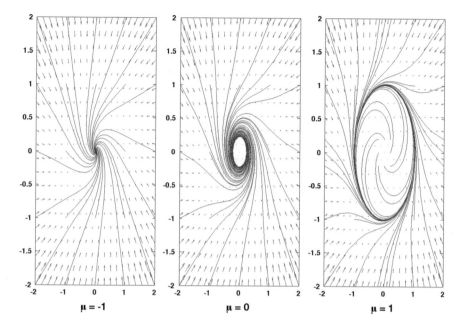

Fig. C.5 Two-dimensional supercritical Hopf bifurcation example

- In subcritical Hopf bifurcation oscillations arise with finite period and finite amplitude. This type of bifurcation is typically associated with the cyclic fold (CF) bifurcation in which stable and unstable periodic orbits are annihilated (or created); with the increase of the bifurcation parameter an unstable spiral becomes stable, while simultaneously the unstable limit-cycle disappears (Fig. C.6)
- At saddle-node-on-invariant-circle bifurcation, a saddle-node collapses and a limit cycle of high period (infinite at the bifurcation point) arises. At the bifurcation point the system has homoclinic orbit.
- Saddle-loop (or homoclinic) bifurcation is somehow similar to SNIC; at bifurcation point, there exists a homoclinic orbit (an orbit which joins a saddle equilibrium point to itself), which then gives origin to a periodic orbit of high period (infinite at the bifurcation point). Note that the "minimal system" exhibiting SL bifurcation has no stable recurrent state (steady state or limit cycle) for $\lambda < \lambda_0$. It is thus not a good candidate for a biological model.

Simple example of the multidimensional system exhibiting supercritical Hopf bifurcation is given by

$$\dot{x} = -x^3 + x(\mu - y^2) - y$$
$$\dot{y} = -y^3 + y(\mu - x^2) + x \qquad\qquad (C.18)$$

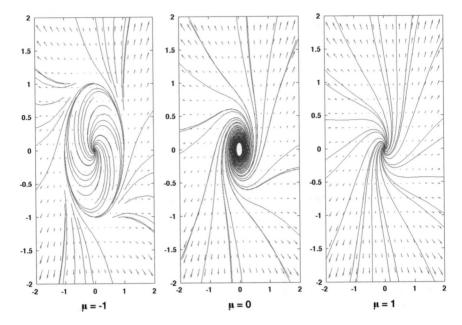

Fig. C.6 Two-dimensional subcritical Hopf bifurcation example

while simple example of the two-dimensional system exhibiting subcritical Hopf bifurcation is given by

$$\dot{x} = x^3 + x(\mu + y^2) - y$$
$$\dot{y} = y^3 + y(\mu + x^2) + x \qquad (C.19)$$

References

1. J. Guckenheimer, P. Holmes, *Nonlinear Oscillations, Dynamical Systems, and Bifurcations of Vector Fields*, 3rd edn. (Springer, New York, 1997), pp. 145–149
2. H. Poincare, L'equilibre d'une masse fluide animee d'un mouvement de rotation. Acta Math. **t.7**, 259–380 (1885)
3. S.N. Rasband, *Chaotic Dynamics of Nonlinear Systems* (Wiley, New York, 1990), pp. 27–30

Appendix D
Numerical Implementation of the Runge–Kutta and Gillespie Methods

D.1 Deterministic Algorithms

Numerical solution of the system of ordinary differential equations can be found by using one of the Runge–Kutta methods. The Runge–Kutta family of methods is based on the same algorithm with different orders and coefficients. Although generally known, we will shortly describe it for clarity of the hybrid, deterministic-stochastic algorithm described later.

Let us consider the following differential equation:

$$\frac{dy}{dt} = f(t, y) \tag{D.1}$$

Assuming we know the initial value of y, the numerical solution of Eq. (D.1) could be represented by the iterative formula:

$$y_{n+1} = y_n + \Delta y \tag{D.2}$$

The difference between various ODE solving algorithms including various Runge–Kutta methods is in determination of Δy. The Runge–Kutta methods family uses a special form of Eq. (D.2):

$$y_{n+1} = y_n + h \sum_{i=1}^{s} b_i k_i \tag{D.3}$$

where

$$k_i = f\left(t_n + c_i h, y_n + h \sum_{j=1}^{s} a_{ij} k_j\right) \tag{D.4}$$

© Springer International Publishing Switzerland 2016
A. Świerniak et al., *System Engineering Approach to Planning Anticancer Therapies*, DOI 10.1007/978-3-319-28095-0

The particular methods differ in the order s and a, b, and c coefficients. The most convenient method to provide these coefficients is via the so-called Butcher table [1].

One of the most commonly used Runge–Kutta methods with constant integration step h is the Runge–Kutta method of order 4. Its Butcher table is given in Table D.1.

According to Eqs. (D.3) and (D.4) it leads to:

$$k_1 = f(t_n, y_n)$$

$$k_2 = f\left(t_n + \frac{h}{2}, y_n + \frac{h}{2}k_1\right)$$

$$k_3 = f\left(t_n + \frac{h}{2}, y_n + \frac{h}{2}k_2\right)$$

$$k_4 = f(t_n + h, y_n + hk_3)$$

$$y_{n+1} = y_n + \frac{h}{6}(k_1 + 2k_2 + 2k_3 + k_4) \tag{D.5}$$

One can notice that when the function is almost constant so i.e.:

$$y_{n+1} - y_n \approx 0, \tag{D.6}$$

the k_i values are approximately equal to zero and the integration step h can be lengthened to accelerate calculations without causing significant inaccuracies in the results. However if the function changes its value rapidly, the h step value has to be shortened to keep the results accurate. It is therefore convenient to develop an algorithm which will adjust the step size to the function variability. Such algorithms are called variable-step methods. The main problem in these methods is to answer the question: when and how to change the integration step size. The easiest way is to perform one step of the chosen algorithm with a very small step size and then to check the difference between the y_n and y_{n+1}. If it is smaller than a given threshold, then the integration step in the next algorithm step should be extended, for example multiplied by 2. If the difference is above another threshold the integration step in the next algorithm step should be shortened, for example divided by 2. Another approach is to use the observation that in the Runge–Kutta methods the accuracy of the y_{n+1} determination grows with the method order. Moreover the difference between the results obtained by using different order methods, called local truncation error, grows with function variability and step size. One can use

Table D.1 Butcher table for the Runge–Kutta method of order 4

0	0	0	0	0
1/2	1/2	0	0	0
1/2	0	1/2	0	0
1	0	0	1	0
	1/6	1/3	1/3	1/6

then, for example, local truncation error for the Runge–Kutta methods of order 4 and 5 to adjust the step size. The methods based on the above observation are commonly used. The Butcher tables for these methods have an additional b^* row of the b coefficients for method of order $n - 1$ and the a, b, and c values in these tables are chosen experimentally. One of the most popular are Bogacki–Shampine method based on Runge–Kutta 2nd and 3rd order, and Dormand–Prince based on Runge–Kutta 4th and 5th order methods.

D.2   Stochastic Algorithms

The most popular algorithm used to simulate the stochastic biochemical reactions and the signaling pathways is the Gillespie algorithm. It is fully stochastic approach resulting in large computational complexity and time consumption. For this reason a number of modifications arose, the main goal of which is to speed up the algorithm without impinging its performance. Base algorithm and its main modifications are described below.

D.2.1   Gillespie Algorithm

For the given time t and state S Gilespie algorithm determines time τ when next stochastic event occurs and reaction μ from all possible reactions which occurs in this time [4].

For that Gillespie defines a probability density function of the reaction $P(\mu, \tau)$,

as the probability that reaction R_μ occurs in the infinitesimal time interval $(t + \tau, t + \tau + d\tau)$, assuming that in the time t the system is in the state $S(t)$

Please notice that the density function $P(\mu, \tau)$ is assumed to be proportional only to $d\tau$. The probability $P(\mu, \tau)\, d\tau$ is calculated using the following formula:

$$P(\mu, \tau)\, d\tau = P_0(\tau)\, P_\mu\, (d\tau) \tag{D.7}$$

where

$$P_\mu(d\tau) = a_\mu\, d\tau \tag{D.8}$$

is the probability that reaction R_μ occurs in the time interval $(t + \tau, t + \tau + d\tau)$ and a_μ is the propensity of the reaction μ.

$P_0(\tau)$ is the probability that for a given time t and state $S(t)$ no reaction will occur in the time period $(t, t + \tau)$. Taking into account the above formulas, the probability that any of the M possible reactions will occur in that time period is equal to

$$P^*(d\tau) = \sum_{\mu=1}^{M} a_\mu \, d\tau = a^* \, d\tau \tag{D.9}$$

The probability that in the time period $d\tau$ no reaction will occur is equal to $1 - a^* \, d\tau$, which gives

$$P_0(\tau + d\tau) = P_0(\tau)(1 - a^* \, d\tau), \tag{D.10}$$

It leads to the following differential equation:

$$\frac{d\,P_0}{d\tau} = -a^* \, P_0, \tag{D.11}$$

solution of which takes the form:

$$P_0(\tau) = e^{-a^* \tau}. \tag{D.12}$$

Based on the above formulas one can use the cumulative distribution to obtain the probability that any of the possible reactions will occur in the time interval $(t + \tau, t + \tau + d\tau)$ summing Eq. (D.7) over all μ propensities:

$$P(\tau) = \sum_{\mu=1}^{M} P(\mu, \tau) \, d\tau = a^* \, e^{-a^* \tau} \tag{D.13}$$

Please notice that Eq. (D.13) gives only the probability that any reaction will occur in the time period $d\tau$ not in the given time. The additional assumption has to be made that the $d\tau$ is negligible. This allows for the precise time calculation when the next reaction will occur.

Probability that the reaction which will occur in the time τ will be μ is a conditional probability $P(\mu|\tau)$ given by the formula:

$$P(\mu|\tau) = \frac{P(\mu, \tau)}{P(\tau)} = \frac{a_\mu \, e^{-a^* \tau}}{a^* \, e^{-a^* \tau}} = \frac{a_\mu}{a^*} \tag{D.14}$$

Equations (D.13) and (D.14) answer the questions when the next reaction will occur and which one it will be.

To define the proper computational algorithm we have to transform the above equations as follows. To determine exact time when the next reaction will occur we have to use the probability density function of Eq. (D.13), which is equal to

$$P(\tau) = a^* \, e^{-a^* \tau}, \tag{D.15}$$

with the cumulative distribution function given by the formula:

$$F(t) = \int_{-\infty}^{t} P(\tau)\,d\tau = a^* \int_0^t e^{-a^*\tau}\,d\tau = 1 - e^{-a^*t} \tag{D.16}$$

With the r_1 number drawn from the uniform distribution on $[0, 1]$, in order to select the time t, so that $F(t) = r_1$ one can determine the t value from the formula:

$$t = F^{-1}(r_1) = \frac{1}{a^*} \ln\left(\frac{1}{1 - r_1}\right). \tag{D.17}$$

Because r_1 is drawn from the uniform distribution on $[0, 1]$, $1 - r_1$ can be replaced by r_1, which gives

$$\tau = \frac{1}{a^*} \ln\left(\frac{1}{r_1}\right) = -\frac{1}{a^*} \ln(r_1) \tag{D.18}$$

The index μ of the reaction which will occur in the given time can be determined by using Eq. (D.14). If we draw r_2 from the uniform distribution on $[0, 1]$, then the occurring reaction index can be determined by using the inequality:

$$\sum_{j=1}^{\mu-1} \frac{a_j}{a^*} \leq r_2 < \sum_{j=1}^{\mu} \frac{a_j}{a^*} \tag{D.19}$$

Now we can define the basic Gillespie computational algorithm:

1. Set $t = 0$ and determine the initial number of all of the molecules involved. Initiate the random number generator.
2. For the given time t and known number of the involved molecules calculate all of the a_μ propensities. Determine $a^* = \sum_{\mu=1}^{M} a_\mu$.
3. Draw two numbers r_1 and r_2 from the uniform distribution on $[0, 1]$. Determine $\tau = (1/a^*) \ln(1/r_1)$ and μ that satisfies the inequality: $\sum_{j=1}^{\mu-1} a_j \leq r_2 a^* < \sum_{j=1}^{\mu} a_j$.
4. Update the molecules number according to the rules of the winning reaction R_μ, advance time t by: $t \rightarrow t + \tau$.
5. Check if the simulation end time T was reached or if all of the $a_\mu = 0$. If yes, then finish the simulations, else go back to point 2.

D.2.2 τ-Leap Method

Despite its simplicity the basic Gillespie algorithm has a significant flaw. After the two random number draws, and time τ as well as reaction index μ determination, only one reaction of the winning type is performed. Let us assume that we have

100,000 molecules A and 100,000 molecules B which in the R_μ reaction bind and create an AB complex. If the reaction μ wins, then in the time $t + \tau$ exactly one A molecule binds one B molecule and creates exactly one AB complex. Assuming that there is no other reaction in the system and that the $A + B \rightarrow AB$ reaction is irreversible, one will need 100,000 steps of the basic Gillespie algorithm to complete the reaction process. This is the source of the large computational load of this algorithm.

In 2001 Gillespie proposed a modification of his original algorithm which he called the τ-leap method [5]. He noticed that when the A and B molecules number is large enough, the system state between t and $t + \tau$ is barely changed. Assuming that the AB complex creation is fast enough, i.e. complex creation rate coefficient in Eq. (D.18) is large, we obtain time τ going to zero. In this situation, after one step of the basic Gillespie algorithm the system is almost in this same state and time as before. As a result the reaction propensities μ are almost the same as before the last step.

Because system transitions to the next states are Markov process realizations, in the given time t and state of the system we do not have to know the exact order of previous transitions to perform the next step. To determine the current state of the system we only have to know how many times each of the possible events occurred from the beginning of the simulation and the initial state. Taking into account the above, Gillespie proposed to omit some of the states during the simulation and then to estimate the number of each reaction which occurs in the omitted time by using the Poisson distribution. The estimation accuracy is based on the previously mentioned assumption of the invariance of the propensity functions. Please notice that when the leap becomes longer, which results for more reaction taking place over the leap time the above assumption becomes less accurately satisfied. This results from the fact that reaction propensities depend on the number of the reacting molecules. For this reason, the leap length τ has to be limited. On the other hand, to maximize the algorithm speed-up effect, the leap length should be as long as possible. The optimal solution will be to find the largest possible τ which will not cause a significant change of any of the μ propensities in the time interval $[t, t + \tau]$:

$$a_\mu(x(t)) \approx \text{const.} \quad \text{for } [t, t + \tau] \tag{D.20}$$

where $x(t)$ is the system state at time t. The above condition is called by Gillespie *the leap condition*. One approach to determine the best τ is to check the post-leap condition:

$$|a_j(x + \lambda) - a_j(x)| \quad \text{for } j = 1, \ldots, M \tag{D.21}$$

Despite its simplicity the above approach is very time consuming. It requires repeating the calculations of the whole system for different τ values as long as the leap condition is satisfied while maximizing the τ. It seems preferable to use the pre-leap condition proposed by Gillespie in [5]. To compute time step, by using the pre-leap check, we define auxiliary variables, $b_{ij}(x)$ and $\xi(x)$, as follows:

$$b_{ij}(x) = \frac{\partial a_j(x)}{\partial x_i} \quad \text{for } j = 1,\ldots,M; i = 1,\ldots,N \tag{D.22}$$

$$\xi(x) = \sum_{j=1}^{M} a_j(x)v_j \tag{D.23}$$

Time τ is chosen using the following formula

$$\tau = \min_{j\in[1,M]} \left[\frac{\epsilon a_0(x)}{|\sum_{i=1}^{N} \xi(x)b_{ij}(x)|} \right] \tag{D.24}$$

where ϵ is an arbitrarily chosen number from $(0, 1)$.

The number k_j of the reaction R_j firings in the time interval $(t, t + \tau)$ can be calculated by using $\lambda = a_j(x)\tau$ as the mean value for reaction R_j and a random draw from the Poisson distribution with intensity λ:

$$k_j = \frac{\lambda^n e^{-\lambda}}{n!} \quad \text{for } n = 0, 1, 2, \ldots \tag{D.25}$$

In short the basic τ-leap algorithm is as follows:

1. for given time t and system state find the τ satisfying the leap condition, i.e. by using Eq. (D.24)
2. using the Poisson distribution draw the numbers of each reaction occurrence in the time range $(t, t + \tau)$,
3. change the system state according to the reaction rules, replace time t by $t = t+\tau$, go back to point 1.

Please notice that if the number of reacting molecules is small enough, such as severed, dozen or tens, each change of their number will have a significant impact on the corresponding propensities, which in the worst case results in the leap length equal to the time to the next reaction. In such cases the τ-leap algorithm becomes equivalent to the basic Gillespie algorithm.

The weakness of the τ-leap algorithm lies in the fact that a species with only few molecules at the given time may be involved in more than one reaction, which should be considered independently. For example, we may have only five molecules of some species that could form complexes and be degraded. Because the number of each reaction firings does not depend on the other reactions it is possible that the total number of the reactions resulting in the decrease of the number of a given species will be greater than the total number of molecules of that species. For example, we may have four complex creations and three degradations between time t and $t + \tau$. This will result in the negative number of molecules of a given species

and algorithm collapse. Solution to this problem is given in the papers of Cao et al.
[2] and Chatterjee et al. [3]. The idea is to limit the possible reaction firings in the
case when the number of reacting molecules is close to 0.

D.2.3   Haseltine–Rawlings Modification

In 2002 Haseltine and Rawlings published their modification of the basic Gillespie
algorithm [6]. According to their postulate, all reactions in the system should be
separated into two groups: fast and slow. According to the mass action law, the
reaction rate depends on the concentrations of the reacting molecules, which could
be extended to the reaction propensities which also depend on these concentrations
and thus on the number of molecules.

Fast reactions are therefore those reactions in which the number of reacting
molecules is large, such as tens or hundreds of thousands. In this case the few
or dozen molecules deviation in the number involved in the reaction since time
t, have negligible impact on the reaction propensities at the given time. The step
size τ before the next reaction occurs is also negligibly small so we can, with high
accuracy, replace the stochastic description with the deterministic one given by the
ordinary differential equation.

In the case when the number of the reacting molecules is small, the reaction
propensities also have small values so these reactions will be slow, with long
τ intervals between successive stochastic events. A small number of reacting
molecules, on the order of few or tens results in a significant deviation in the
reaction propensities in the cases when barely few of them occurred. Because of
that, the deterministic approximation of the stochastic processes will be loaded with
significant error, and in this case it should not be used.

Based on the above insight Haseltine and Rawlings proposed computational
algorithm which can be summarized in the following steps:

1. for a given system separate all reactions into two groups: slow and fast. Describe
   the fast ones by using the ordinary differential equations and the slow by using
   the reaction rules.
2. for a given time t and known number of molecules calculate all reactions
   propensities $a_\mu(t)$ for the slow reactions and then total propensity function:

$$a^*(t) = \sum_{\mu=1}^{M} a_\mu(t) \tag{D.26}$$

3. draw two numbers r_1 and r_2 from the uniform distribution on $[0, 1]$

4. evaluate the system of the ODEs, e.g. by using Eq. (D.5) for the fast reaction from time t to time $t + \tau$ such that

$$\log(r_1) + \int_t^{t+\tau} a^*(s)ds = 0. \tag{D.27}$$

5. by using the inequality:

$$\sum_{j=1}^{\mu-1} a_j(t + \tau) \leq r_2 a^*(t + \tau) < \sum_{j=1}^{\mu} a_j(t + \tau) \tag{D.28}$$

where $a_j(t + \tau)$ stand for the individual reactions propensities, determine index μ of the slow reaction which will occur at the time $t + \tau$

6. change the system state according to the winning reaction rules. If the stop condition is not met, replace time $t \to t + \tau$ and return to point 2.

The above algorithm allows accelerating significantly the simulation, while maintaining the stochasticity of the system. From the practical point of view, point 4 of the above algorithm could be problematic. This is due to the fact that reaction propensities a_μ and total propensity a^* are not constant between stochastic "firings" as in the Gillespie algorithm but are functions of reacting molecules number and therefore time. Because of that it is impossible to calculate the time $t + \tau$ to which the ODE should be evaluated. The above problem can be solved in two ways:

1. ODE evaluation to the time T large enough, that for every possible realization of the ODE, condition (D.27) will be satisfied. Knowing ODE trajectories we know how the reaction propensities change in time and we can find time $t + \tau$.
2. Regular sampling of the ODE system every Δt to determine if the stochastic reaction occurs in the period time from the last check-time to current check-time.

The first approach is simpler. It allows the use of publicly available ODE solvers. The disadvantage of this approach is the need of assuming very large T comparing to the expected time to the next stochastic reaction occurrence to maintain the correctness of the obtained results. Please notice that after $t + \tau$ determination and execution of the winning reaction, the ODE realization from range $(t + \tau, T)$ becomes invalid and should be forgotten. Because usually $t+\tau \ll T$ a large amount of computing power is wasted on unnecessary calculations, which substantially increases the simulation time.

The second approach provides faster simulation but requires its own ODE solver implementation, e.g., according to Eq. (D.5). In this case the integral from Eq. (D.27) is calculated step by step, as next y_{n+1} is determined, to check if the condition (D.27) is satisfied and therefore $t + \tau$ found. In this method, some inaccuracies may result from the sampling time, so it is necessary to assume small enough integration step h, i.e. one or two order smaller than the time expected to the next reaction.

References

1. J.C. Butcher, *Numerical Methods for Ordinary Differential Equations* (John Wiley & Sons, New York, 2008)
2. Y. Cao, D.T. Gillespie, L. Petzold, Efficient Stepsize selection for the Tau-leaping method. J. Chem. Phys. **124**, 044109 (2006)
3. A. Chatterjee, D.G. Vlachos, M.A. Katsoulakis, Binomial distribution based τ-leap accelerated stochastic simulation. J. Chem. Phys. **122**, 024112 (2005)
4. D.T. Gillespie, A general method for numerically simulating the stochastic time evolution of coupled chemical reactions. J. Comput. Phys. **22**, 403–434 (1976)
5. D.T. Gillespie, Approximate accelerated stochastic simulation of chemically reacting systems. J. Chem. Phys. **115**, 1716–1733 (2001)
6. E.L. Haseltine, J.B. Rawlings, Approximate simulation of coupled fast and slow reactions for stochastic chemical kinetics. J. Chem. Phys. **117**, 6959–6969 (2002)

Index

Adenoma, 128
Advection-reaction-diffusion equation, 89
Angiogenesis, 61, 62, 85–87, 98, 127, 176
Antiangiogenic therapy, 4, 61, 69, 72, 176
Apoptosis, 22, 35, 58, 107, 140, 154, 163, 188
Asymptotic behavior, 70–72, 111, 121
Asymptotic expansion, 42
Asymptotic properties, 37
Avascular growth, 61, 94, 98

Bifurcation, 79, 153, 156, 158, 183, 213
Bone marrow, 33, 44, 105, 114
Branching process, 106, 111, 115, 116, 143
Breast cancer, 21, 37, 103, 163, 175

Carcinogenesis, 1, 6, 115, 158, 174
Cell cycle, 5, 9, 56, 103, 108, 119, 130, 165, 174
Checkpoint, 11, 19, 57, 123
Chemotaxis, 86, 99, 127
Chemotherapy, 2, 11, 13, 17, 33, 61, 69, 72, 77, 97, 104, 117, 118, 128, 129, 165
Combined therapy, 32, 63, 69, 70, 72
Control
    bang-bang, 26, 27, 33, 47, 69, 74, 210
    optimal, 24, 41, 47, 68, 75, 207
    singular, 25, 27, 47, 68, 74, 209
Control system
    closed-loop, 3, 24
    open-loop, 3
Controllability, 70, 199
    global, 71
    local, 70, 71, 204

Cytostatic agents, 14, 37, 104, 120
Cytotoxic agents, 5, 12, 18, 21, 34, 69, 72, 104, 120, 165

Degradation, 30, 35, 89, 100, 143, 144, 151, 152, 161, 178, 188
Desmoplastic tumor, 92
Differentiation, 22, 106–108, 112
DNA damage, 30, 35, 103, 140, 157, 158
DNA repair, 35, 144, 158, 165
Drug resistance, 2, 5, 35, 36, 61, 69, 73, 104, 117, 121, 175
Drug sensitivity, 35

Equilibrium, 12, 110, 113, 183, 185, 199, 214
Equilibrium point, 28, 59, 64, 67, 69, 79, 156, 200, 218

Feedback, 1, 21, 78, 108–110, 113, 125, 139
    negative, 3, 61, 107, 140, 145, 151, 153, 156, 158, 163
    positive, 107, 140, 145, 152, 153, 156, 158, 163

Gene amplification, 36, 46, 107
Gene therapy, 75
Genotype, 93
Gillespie algorithm, 143, 225
Glioblastoma, 79
Granulocyte, 107, 113
Growth factors, 10, 90, 92, 98, 105

© Springer International Publishing Switzerland 2016
A. Świerniak et al., *System Engineering Approach to Planning
Anticancer Therapies*, DOI 10.1007/978-3-319-28095-0

Printed in the United States
By Bookmasters